Democracy, Dialogue, and Environmental Disputes

Democracy, Dialogue, and Environmental Disputes
The Contested Languages of Social Regulation

Bruce A. Williams and Albert R. Matheny

Yale University Press / New Haven and London

Set in Ehrhardt type by The Composing Room of Michigan, Inc., Grand Rapids, Michigan.

Printed in the United States of America by BookCrafters, Inc., Chelsea, Michigan.

Library of Congress Cataloging-in-Publication Data

Williams, Bruce Alan.

    Democracy, dialogue, and environmental disputes : the contested
languages of social regulation / Bruce A. Williams and Albert R.
Matheny.

      p.   cm.

    Includes bibliographical references and index.

    ISBN 0-300-06241-9 (alk. paper)

    1. Policy sciences. 2. Communication in politics—United States.
3. Delegated legislation—United States. 4. Environmental policy—
United States. 5. Hazardous wastes—Law and legislation—United
States. I. Matheny, Albert R. II. Title.

H97.W544 1995

363.7—dc20   95-7650

                                                  CIP

A catalogue record for this book is available from the British Library.

The paper in this book meets the guidelines for permanence and durability of the Committee on Production Guidelines for Book Longevity of the Council on Library Resources.

10 9 8 7 6 5 4 3 2 1

This book is lovingly dedicated to our wives, Andrea Press and Jane Adair, and to our future, Jessie and Al.

Contents

Acknowledgments

We started graduate school together two decades ago at the University of Minnesota. The days we spent together there remain very special in our memories. This is true for just about anyone who has pursued an academic career; but for many, those times become only memories, and they fade away. The great friendships of graduate school may never be matched again in a lifetime, yet they can easily die, as careers divide and conquer our professional and private lives. If it weren't for this book, the same might have happened to us. We hatched some of the ideas for it back in 1978 during our last summer at Minnesota. While those ideas have long since been unrecognizably absorbed into the theoretical portions of the book, we have sometimes surprised ourselves by discovering threads of an argument that lead consistently back to those early discussions.

Figuring out the regulatory process took us longer than we expected, to say the least. As our careers and families evolved, there were many distractions; but our evolving intellectual interests, the book, and the friendship reinforced one another and kept us together over a decade that otherwise would have pulled us apart. Following time spent in the field together, a pattern of writing and rewriting gradually emerged and endured over many years and sometimes intercontinental distances. Indeed, an indication of how long we have spent on this project is that our collaboration has gone through several generations of communications technology: regular mail, long distance telephone, express mail, and finally electronic mail (over which we constructed the final version of the book, one of us in Ann Arbor and the other in Utrecht, Netherlands).

In any project that takes this long to finish, many debts are accumulated. Here we can only scratch the surface in acknowledging those debts by recognizing some of those who helped us along the way. Michael Delli Carpini, Bob Lake, Robert Paehlke, Andrea Press, Barry Rabe, Tony Rosenbaum, and Carmen Sirianni all read and commented on the manuscript. We have benefited from ongoing interactions with a group of scholars who have been concerned with the issue of hazardous waste regulation and the problems of democratic participation in policy-making.

Among those in this group from whom we have learned a great deal are Anne Bowman, Chuck Davis, John Dryzek, Christine Harrington, Michael Kraft, Jim Lester, Dan Mazmanian, Sidney Milkis, and Kent Portney. The University of Florida and Florida Atlantic University generously supported our travel early in the interviewing phases of the project. Invitations to present our ideas at Pennsylvania State University and Clark University helped our thinking at especially opportune times.

Bruce Williams would like to acknowledge two special debts. One is to Andrea Press, who has been colleague, companion, and wife. She took time off from an incredibly hectic and successful academic career of her own to read and comment on countless drafts of our manuscript. More important, Andrea had more faith in this project than we had, and perhaps more faith than was wise, but without her you would not be reading this book now. A second debt is to Jack Meiland of the University of Michigan, who contributed to this project in many obvious and not so obvious ways as a friend, an intellectual kindred spirit, and a model of what it means to be an academic, in the very best sense of the word. Albert Matheny especially thanks his wife, Jane Adair, for whom the connection between dialogue and community comes naturally, and Les Thiele of the University of Florida, whose friendship and example as well as his thinking about political community and the environment have influenced our conclusions.

This book would not have been possible without the help of the many people in New Jersey, Ohio, and Florida who graciously agreed to be interviewed and reinterviewed over a ten-year period. So many people took time off from their busy schedules in government, business, and various interest groups that it is impossible to name them all; nevertheless, we are grateful to them all. Lois Gibbs of the Citizens Clearinghouse for Hazardous Waste and the Ironbound Community of Newark, New Jersey, both on the front lines of the grassroots regulatory struggle, were especially helpful. Their words and deeds contributed profoundly to the way we think about regulation and citizenship.

Following the conception of our project as a book, we have worked under the watchful eye of John Covell at Yale University Press. His support for the particularly long gestation of our efforts, from initial contract to final editing, has been extraordinary. Lawrence Kenney earned our deep appreciation for his fine job of editing our manuscript. Of course, all the support we have received over the years cannot absolve us of our own responsibility for the final product. However, if you find something you don't like, it was probably Bruce's idea.

Abbreviations

BFI	Browning-Ferris Industries (multinational waste management company)
CAP	Community Action Program
CCHW	Citizens' Clearinghouse for Hazardous Wastes, Inc.
CECOS	Chemical and Environmental Conservation Systems (national company operating in Ohio)
CER	Clermont Environmental Reclamation Company (Ohio)
CERCLA	Comprehensive Environmental Response, Compensation, and Liability Act
CFACT	Committee for a Constructive Tomorrow (corporate-sponsored "grassroots" group)
COSH	Committee on Occupational Safety and Health
DEP	Department of Environmental Protection (New Jersey)
DER	Department of Environmental Resources (Florida)
ECRA	Environmental Cleanup Responsibility Act (New Jersey)
EPA	Environmental Protection Agency
ERC	Environmental Regulation Commission (Florida)
ERRIS	Emergency and Remedial Response Information System
GAO	General Accounting Office
GNP	Gross National Product
GSP	Gross State Product
HSWA	Hazardous and Solid Waste Amendments
HWAC	Hazardous Waste Advisory Council (New Jersey)
HWDF	Hazardous Waste Disposal Facility
HWFAB	Hazardous Waste Facilities Approval Board (Ohio)
HWFB	Hazardous Waste Facilities Board (Ohio)

HWFSC	Hazardous Waste Facilities Siting Commission (New Jersey)
HWTC	Hazardous Waste Treatment Council (industry lobbying group)
I-CARE	Independent Citizens Associated for Reclaiming the Environment (Ohio)
LEPC	Local Emergency Planning Committee
LSC	Legal Services Corporation
NIMBY	Not-in-My-Back-Yard
NIOSH	National Institute of Occupational Safety and Health
NPL	National Priority List
OEPA	Ohio Environmental Protection Agency
OMB	Office of Management and Budget
OSHA	Occupational Safety and Health Administration
RCRA	Resource Conservation and Recovery Act
SARA	Superfund Amendment and Reauthorization Act
SLAPP	Strategic Lawsuit against Public Participation
TSCA	Toxic Substances Control Act
UCAP	United Citizens against Pollution (Florida)
VOICE	Voting Ohioans Initiating a Clean Environment
WQAA	Water Quality Assurance Act (Florida)
WQATF	Water Quality Assurance Trust Fund (Florida)

PART I / The Competing Languages of Social Regulation

1 / Introduction

On April 21, 1980, the tenth anniversary of Earth Day, thousands of drums of hazardous wastes stored at Chemical Control Corporation in Elizabeth, New Jersey, caught fire and exploded. Only luck prevented a monumental disaster as winds blew the fire's huge toxic cloud out to sea rather than over Manhattan. The visual drama of the event, the possibility of having to evacuate New York City, and later media coverage linking organized crime to Chemical Control and documenting government mishandling of the site kept the incident in the media and in the public eye for some time. A more dramatic example of the problems of improper regulation and storage of hazardous wastes is difficult to imagine. Responding to the disaster in Elizabeth as well as to other, less dramatic events, the New Jersey legislature decided that building secure disposal sites was central to regulating hazardous substances and, in 1981, established a new procedure for siting such facilities.

Successful siting of these facilities seemed certain in New Jersey. A national leader in the area, the state had already passed much legislation that later served as a model for other states and even the federal government. The serious nature of the problem, evidenced by events like the Elizabeth explosion, kept the issue on the public agenda and marshaled support for a solution. An active and established environmental movement in the state channeled public outrage into political pressure on lawmakers, and there was little industry opposition to the legislation. Indeed, the procedure finally adopted won the support of virtually all established groups involved in hazardous waste regulation specifically and in environmental issues generally.

Many observers viewed the law as a model for the nation because it balanced a reliance on technical and scientific criteria to determine sites with the inclusion of affected interests in the decision-making process.[1] One important policymaker on the board responsible for determining the safest disposal sites said, "The New Jersey siting plan is the only way to get siting done. First, you have to accept that there is a risk, then look for ways to minimize that risk in case of an accident."[2] In addition to developing technical criteria for risk reduction, the legislation recognized the need to include a broad array of groups involved in environmental policy-making. The widespread support for the plan was emphasized by Diana Graves, head of the state's Sierra Club: "Because there was universal recognition in New Jersey that we needed treatment facilities to better handle hazardous waste, there was agreement on the provisions that needed to go into the act. There was support for the act throughout the state."[3]

In spite of the enthusiasm shown for the legislation and the broad support for the siting process it created, New Jersey in 1990 was no closer to building a facility than it had been in 1980 when the Elizabeth fire occurred. Instead, the emergence of grassroots groups in communities with existing hazardous waste problems and in those selected for new sites had derailed the process. These groups, arguing that their interests were being ignored, have created a rift in the state's environmental movement, causing problems for government officials by raising questions that were overlooked in the creation of regulations. The arguments of such groups are nicely articulated by a member of Newark's Ironbound Community:

> They [the state-level environmentalists] are out of touch. We are slowly building alternatives to raise consciousness. There is a new environmental movement, and people—local people—are fighting to effect enforcement rather than simply symbolic regulation. The reason is simple. If hazardous waste is not an "act of God," then ask yourself, "Who benefits"? Industry. Government won't stop it; so we are left with the responsibility.[4]

New Jersey's experience was similar to that of other cases we studied, as all across the country states and localities struggled to regulate the production and disposal of hazardous wastes.[5] In case after case, grassroots groups stalled the policy process. This phenomenon has been called, pejoratively, the Not-In-My-Back-Yard (NIMBY) movement. Most generally, it refers to the mobilization of citizens, especially in neighborhoods or communities, to resist the siting of unpopular facilities (hazardous waste facilities, low-level nuclear waste facilities, and prisons as well as certain types of factories and other commercial ventures). We believe that such struggles are significant because they are concrete manifestations of American society's attempts to regulate the costs of economic growth and technological development. Why stalemate is so often the result of these disputes is a central concern of this book.

Many specific questions are raised by these struggles. First, why does policy break down despite general agreement over both the severity of the problems posed by threats like hazardous wastes and the need to manage such problems safely? Second, why do policymakers and other participants in the policy process so routinely fail to anticipate the objections raised by newly formed NIMBY groups like the Ironbound Community? Third, why are these challenges so successful at stalling the policy process? Fourth, is it possible to design policy-making processes capable of avoiding such misunderstandings and omissions?

Posing these questions is standard for public policy studies undertaken by political scientists. However, answering them satisfactorily requires a broader view than that traditionally employed by students of the policy process. Narrowly focused policy studies miss the degree to which specific controversies reflect more general and unresolved questions about a democratic political system's ability to weigh the costs and benefits of progress. In fact, we chose to study disputes over hazardous waste regulation and disposal precisely because they illuminate the connection between such general questions and the dynamics of specific policy disputes. Regulatory issues at the state and local levels, like hazardous waste regulation (which we study in chapter 6) or right-to-know legislation (which we study in chapter 7), require political communities to confront difficult issues: balancing the trade-offs between material consumption and environmental degradation; distributing the risks, benefits, and uncertainties of technological developments; and, fundamentally, providing citizens with the information needed to debate such issues and then translating their disparate political demands into public policy.

Because they involve the interference of government in private markets, all regulatory policies are potentially controversial. However, some types of regulatory policy are more controversial than others. In particular, the distinction between economic and social regulation is critical. Economic regulation concerns itself primarily, although not exclusively, with market failures resulting from natural monopoly.[6] There is a well-developed and widely accepted body of economic analysis that explores the shortcomings of much economic regulation. The deregulation of industries once thought to be natural monopolies is evidence of the persuasive power of this literature.[7]

There is much less agreement about social regulation, the broad policy arena within which fall the controversies we studied. Whereas economic regulation addresses natural monopoly, social regulation deals with "the externalities and impacts of economic activity."[8] In areas like environmental protection, health care reform, occupational safety, and consumer protection, these externalities and impacts often include difficult to define quality of life issues, requiring regulators to "affect the economy in nonmarket dimensions."[9] This makes such policies inevitably controversial because of disagreements about how to measure the value of the resources at issue.

For example, price closely captures the full value of an airline ticket, a train ride, or a kilowatt hour of electricity: the challenge for economic regulation becomes accurately attaching price tags to such commodities when markets fail to do so. On the other hand, price is less appropriate for capturing the full value of a healthy environment or of avoiding cancer. Although their value certainly has a component that is captured by money, they are not traded on the market and estimating their price becomes a political rather than an economic determination. Further, there is moral resistance to attaching any price tag to such items that lie outside "the domain of dollars."[10] Hence, establishing a "price" for human life or clean water in regulatory calculations remains controversial.[11] Social regulation, then, raises fundamental questions about the relations between technological progress, capitalism, and democracy.

In short, social regulation is a policy arena within which the structures of the state, the private market, and democracy collide. The historical dimensions of this collision are central to understanding specific policy conflicts: local disputes about social regulation are part of an ongoing debate about the contours of citizenship in the United States. In this book, we argue that the boundaries of this debate are defined by the competing languages or discourses that have developed in America for discussing such issues. These languages constitute the competing vocabularies available for participants to articulate their perspectives on and interests in the issues raised by social regulation. Analyzing these languages allows us to understand how their unstated and conflicting assumptions determine what contending interests can and cannot communicate to each other. Our study of specific issues as they arise in local communities, like the siting controversy in New Jersey described above, leads us to consider how these languages are used to constitute citizenship, participation, and democracy in public policies, especially those with profound implications for life in a modern, technologically sophisticated society.

Overview

In part I of this book we lay out our theoretical argument. Chapter 2 identifies three specific languages that have evolved for discussing and debating social regulatory policies. These languages—the pluralist, the managerial, and the communitarian—have deep roots in American political history. The first two are commonly used by governmental officials and other parties routinely included in regulatory decision making: in New Jersey, they were the discourses adopted by state policymakers, industry groups, and state-level environmental groups. The communitarian language challenges policy outcomes and is used by groups (especially at the local level) that are excluded from decision making, but that are nevertheless called upon to bear the costs of social regulation. This was the language of the Ironbound Community and other grassroots activists.

No one of these languages in itself provides an adequate vocabulary for addressing the full range of issues raised by social regulation. A detailed exploration of these limits is the task of chapter 2. Close analysis is required because the inadequacies of each discourse are not at all self-evident. If they were, policymakers, citizens, and scholars would resist more strongly their tendency to couch their arguments in one language while ignoring or discounting claims made in terms of the other languages. The inadequacies of each language lie in the underlying assumptions each makes with respect to the central and contested concepts (for example, science, technology, participation, and so forth) involved in the type of social regulation that concerns us in the book. In the policy process, disputes intensify when antagonists fail to find a common language with which to express their differences. We argue that the absence of political institutions and policy-making procedures capable of fostering common and public dialogue accounts for the stalemate so common in specific social regulatory controversies.

This argument provides a way of grasping more fully the specific policy disputes we address later in the book. But, more important, it is rich with significance for the general study of politics. Political scientists have been slow to understand the implications of what is often labeled postmodernist scholarship, which in many other disciplines points to the difficulty or even impossibility of reducing complex social phenomena to a single discourse or narrative. Yet the insights of this work are essential for appreciating the dynamics of policy disputes as they play out in contemporary American politics. Following from this, a central thesis of our book is that the issues posed by social regulation can best be understood by viewing policy-making as neither a managerial, pluralist, nor communitarian process, but rather as an ongoing, continuously changing public dialogue among citizens, organized interests, and policymakers.

This thesis forms the basis for our dialogic model of policy-making, developed in chapter 3. We use the dialogic model both to analyze actual policy disputes and to raise questions about what constitutes democratic citizenship in a modern, technologically sophisticated society. While this model draws upon the insights of the communitarian discourse (in ways that most policy analysis does not), it rejects the assumption that any single language can comprehend the dynamics of many difficult policy issues. Our model assumes that political truth in a democratic society must always be negotiated and can emerge only from an open, ongoing engagement of all these languages. Yet actual policy debates usually exclude certain languages or ignore the underlying and competing assumptions of the languages used. Existing political institutions fail to house a dialogue among contending interests capable of clarifying, let alone resolving, the disparate assumptions of the differing discourses used for debate.

Chapter 4 places the languages of regulatory legitimacy and the dialogic model itself within the structural reality of American politics, specifically the intermingled political economies of federal, state, and local government. Here we address the issue of federalism and the complications it introduces for implementing such redistributive regulatory policies as those associated with hazardous waste management.

Part II of the book applies the theoretical arguments of part I. In these chapters we explore the ways in which the three languages of social regulation are used in the development of hazardous waste regulation at the national, state, and local levels. We trace how the underlying assumptions of the three competing discourses structure policy disputes in ways unrecognized by participants in these struggles. Using our dialogic model, we also investigate how political institutions and actors might begin to recognize these limits. Chapter 5 provides an overview of the development of hazardous waste regulation at the national and state levels. Chapter 6 reports on our case studies of local hazardous waste disputes in New Jersey, Ohio, and Florida. Chapter 7 examines the attempts of grassroots environmental groups to transcend the limitations of the communitarian language and develop a more contextual understanding of their local struggles. Chapter 8 offers our concluding thoughts about the reform of regulatory politics in America and how it can make policy dialogue a real possibility in the future.

Why Focus on Language? Why Focus on Localities?

As the outline of our argument suggests, this book emphasizes two dimensions generally overlooked in studies of social regulation: a focus on competing languages used in policy-making and attention to the implications of social regulatory issues for local communities. In our view, both dimensions are central for understanding the stalemate that characterizes so much social regulation in particular and public policy generally.

Concern over the use of competing languages in political communities has ancient roots. Thucydides argued that societies fall apart when people disagree over the meaning of basic terms.[12] We argue that, if not society, then certainly public policy can fall apart when participants fail to understand that words mean different things to contending parties. As William E. Connolly argues, certain concepts are likely to generate a variety of meanings by their very nature: "When disagreement does not simply reflect different readings of evidence within a fully shared system of concepts, we can say that a conceptual dispute has arisen. When the concept involved is *appraisive* in that the state of affairs it describes is a valued achievement, when the practice described is *internally complex* in that its characterization involves references to several dimensions, and when the agreed and contested rules of application are relatively *open*, enabling parties to interpret even those shared rules differently as new and unforeseen situations arise, then the concept in question is an 'essentially contested concept.' "[13] Connolly argues that terms like *democracy, politics, public interest, power, responsibility*, and *freedom* are examples of essentially contested concepts. Most of the concepts framing social regulation are essentially contestable. Each of the languages used to debate these policies provides an interconnected and competing set of definitions for these concepts. A central argument of this book is that many of the disputes over social regulation, especially at the local level, prove intractable because of the absence of a public language adequate for expressing the differing assumptions made by these competing languages. We argue that a democratic politics of social regulation would provide a public sphere for more open debate over the definition of contested terms.

Our call for a sharper focus on discourse and the development of political institutions able to account for the impossibility of reducing competing languages to a single viewpoint or language echoes the arguments of many postmodern scholars. There is a rich and growing literature in a variety of disciplines that attempts to address the implications of the postmodern "incredulity toward metanarratives" and the connected claim that all attempts at developing such overarching discourses are simply exercises of power.[14] That is, postmodern theorists claim that the establishing of single languages or vocabularies for understanding the world inevitably distorts communication by suppressing alternative perspectives. We believe that the insights of the debates swirling around what Richard J. Bernstein calls the "constellation of modernism / postmodernism" have direct relevance for the study of public policy.[15] Given the complexity of political struggles, the competing claims of affected interests, and the difficulty of balancing democratic procedures with the use of power and force in government, political scientists have much to learn from and much to offer to this debate. This book is a first step in that direction.

One reason that the insights of postmodernism have had so little impact on political scientists may be that postmodernism's focus on concepts like discourse, texts, vocabularies, and so forth is sometimes taken to deny the existence of political, economic, and social structure. So, the arguments of scholars as diverse as Jacques Derrida and Michel Foucault are ignored because they seem to treat everything as if it were just a text or just a discourse with no connection to the underlying economic and political structures of power that are central concerns of political scientists. Although this conclusion is reinforced by the often dense and difficult writings of many postmodern writers, it is unwarranted. For instance, Foucault, a pioneer in developing a theory of discourse, was always interested in the ways in which various discourses reflect and influence both individuals and institutions. For him, "all social institutions and practices, including the law, the political system, the church, the family, the media . . . are located in and structured by a particular discursive field."[16] This concept of a discursive field stresses the competing "ways of giving meaning to the world and of organizing social institutions and processes."[17] In this way it emphasizes the relations between language, social institutions, and power.

Most definitions of *discourse* used in this literature recognize the connection between language and social, political, and economic structure.[18] John Fiske, for example, defines the term as "a language or system of representation that has developed socially in order to make and circulate a coherent set of meanings about an important topic area. These meanings serve the interests of that section of society within which the discourse originates and which works ideologically to naturalize those meanings into common sense."[19] It is in this sense that we use the terms *language* and *discourse* when we explore the competing languages of social regulation.

In short, our concern with the languages used to debate regulatory policies does not deny the significance of political and economic structure. Neither does it seek to overturn other ways of looking at the political world more familiar to political scientists. Rather, it suggests that the explicit articulation of the structural conflicts involved in the policies we study is limited and shaped by the languages used and the misunderstandings that flow from their differences. Identifying these languages as manifestations of enduring and competing discourses in American politics connects debates over social regulation to broader debates over the shape of capitalism and democracy in the United States. Disputes over social regulation remain difficult to resolve because they are extensions of unresolved deeper conflicts in American political economy.

The second dimension of our study relates to the ways in which social regulation operates when it imposes costs and / or benefits on geographically bounded communities. Whereas most scholars who examine the dynamics of social regulation emphasize national policy-making, we concentrate on states and localities.[20] By studying the ways in which citizens respond to policies that impose significant costs and, less frequently, benefits on their communities the investigator can clarify many issues of participation and democracy obscured by a focus on national policy-making. Most analysts bemoan a decline in political participation of all sorts, but we found that citizens responded vigorously, though often in nontraditional ways, to threats to their communities posed by social regulatory policies: it was at this level that the communitarian language was most clearly expressed. Citizens may be suspicious of and reluctant to engage politics and politicians at the federal or state level, but they are often willing to act politically when public policies have a direct impact on their communities.[21] Building upon this willingness may be a pathway to reconstructing citizenship in the United States.

2 / The Search for the Public Interest in Three Languages: Regulation and Democracy in the United States

The Managerial Language

Charles Beard, writing in 1931, expresses the essence of what we call the managerial language and its assumptions about democracy in modern society: "[Technological revolution] 'has emphasized as never before the role of government as a stabilizer of civilization,' [and confronted it with bewildering complexities. But] 'technology has brought with it a procedure helpful in solving the problems it has created; namely, the scientific method,' which 'promises to work a revolution in politics. . . . It punctures classical oratory—conservative as well as radical. . . . Disputes about democracy, therefore, creak with rust.' "[1]

Many beliefs about regulation that are taken for granted today emerged during the Progressive era, when the organizational society was being built and, as Beard's writing so eloquently testifies, faith in science and technology as remedies for the inadequacies of participatory democracy was at its height. These assumptions constitute the managerial language, a discourse framed by Progressive ideas about the public interest, participation, and democracy. Still used by policymakers to understand and debate regulatory policies, including social regulation, this language emphasizes the technical aspects of regulatory policy and identifies the need for expertise as both a structural barrier to and substitute for democratic participation.[2]

The managerial language remains especially important because the terms used by Progressive reformers have become deeply embedded in the culture of bureaucratic agencies charged with regulatory policy. Because they have become the rules that guide institutional behavior in American regulatory policy, Progressive-era beliefs provide one set of solutions still applied to problems in American regulatory policy.[3]

The Progressive movement of the early twentieth century produced a body of thought central to an understanding of the response of American political institutions to the emergence of modern capitalism. The four decades straddling 1900 saw the rationalization and organization of American life as a national economy emerged, an economy characterized by large, bureaucratically organized corporations competing in oligopolistic markets.[4] These developments posed a challenge to Progressive political thinkers: how to reconcile prosperity with the clear violations of American individualism that modern capitalism seemed to entail? In the words of Richard Hofstadter, "[Progressives] were trying . . . to keep the benefits of the emerging organization of life and yet to retain the scheme of individualistic values that this organization was destroying."[5] Further, Progressives represented the modern middle class, which stood between (and viewed with suspicion) the barons of big business and the poor and restless masses. They sought to use government power to curb the excesses of corporate capitalism without turning that power over to mass democracy, which might undermine the very economic order they sought to save.

The Progressives' solution, especially in regulatory policy, was to rely on trained experts located within bureaucratic organizations committed to the discovery of an objective public interest. Just as the engineering wonders of the age applied science to the physical world and just as Frederick Taylor had developed a science of management, Progressives held that it was possible to develop a science of politics to guide government actions. This science of politics, applied to the administration of regulatory policy, would transcend the narrow, suspect self-interest of various groups and serve the common good. Progressive rhetoric held that scientific expertise could produce policies in the public interest and overcome the inadequacies of democratic decision making.[6]

Such ideas found expression in the separation of politics from administration advocated by Progressive reformers. Elected politicians in a democracy were to set the broad goals of public policy, but the carrying out of such policies was to be a technical matter calling for the application of specialized expertise. Politically neutral administrators trained in the supposed science of politics would be able to carry out policies in the most efficient manner possible and serve the public interest to a much greater extent than uninformed or corrupt elected politicians.

The Progressive era decisively transformed the discourse of American democracy by shifting discussion away from concern over participation in democratic politics to a focus upon neutral, scientific criteria for judging public policy. To quote Russell Hanson, "These criteria—efficiency and dependability—were derived from ostensibly scientific analyses of organizational behavior and economic development that drew heavily upon an evolutionary understanding of American society and its possibilities."[7] Consistent with such beliefs, economic regulation was considered a technical issue, one better addressed by experts than by self-interested citizens or elected politicians. These attitudes long allowed regulators and students of regulation to avoid many troubling issues concerning the relation between democracy and policy-making.

Evolution of the Managerial Language and the Prophets of Regulation

Of course languages are not fixed. Rather, they evolve within certain parameters as the political world they help to interpret and define also changes. Thomas K. McCraw's *Prophets of Regulation* illustrates the development of the managerial discourse throughout the history of regulatory policy-making in the United States. In this prize-winning study, McCraw chronicles four regulators and the institutions they helped design: Charles Francis Adams, a central figure in nineteenth-century attempts to regulate the railroads; Louis D. Brandeis, a supporter of antitrust legislation and influential in the creation of the Federal Trade Commission; James E. Landis, the architect of the Securities and Exchange Commission and one of the most important regulators of the period 1930–60; Alfred E. Kahn, chairman of the New York State Public Service Commission and the Civil Aeronautics Board, a principal proponent of economic analysis as a guide for regulatory policy. McCraw makes clear the Progressive influence on regulatory thought and the extent to which regulatory policy has been shaped by the assumptions of the managerial discourse. Except for Brandeis, whom we discuss below, the subjects of McCraw's study all saw the public interest as something that could be discovered by neutral, scientific experts (even though they disagreed over the appropriate training for such experts) located within bureaucratically organized regulatory agencies. Issues of due process and democratic participation, central to the pluralist and communitarian languages, were neglected by these founders.

Charles Francis Adams was convinced that information and education were the keys to successful regulation. He believed that regulation was a technical matter and that, with the proper information, policymakers could discover the one best policy that would further the interests of all. In the case of railroads, this policy would rationalize the industry and lead to both lowered rates and higher profits. Consistent with Progressive beliefs, this one best policy—discoverable by scientific experts—would transcend the petty political squabbles of self-interested groups and serve the general public interest.

Adams's views about the technical nature of regulatory policy-making allowed him to avoid considering due process, participation, or democratic values. If there were a best policy that would serve the general public interest, then policymakers needed only to gather the information required to discover that policy. Allowing policy to be dominated by the narrow interests of the railroads themselves, consumers, railroad laborers, or any other group could only frustrate the public interest. Indeed, it is not far from Adams's view to the notion that participatory, democratic methods of decision making are inimical to good policy-making. McCraw notes that Adams never saw the conflicts of interest that were developing in industrial America as legitimate. This position was characteristic of Progressives in general: "Such 'disharmony' was inconsistent with the reformers' hope for progress toward a society in which class and ethnic differences had been transcended via the application of neutral, scientific principles of management."[8] Further, reflecting Progressive tenets about education being a prerequisite for mass political participation, Adams believed that until there was widespread acceptance of the appropriate analytic techniques and modes of information on the part of the public and interest groups, the democratic political process would likely produce bad regulatory policies. The managerial language's conception of education as a way of achieving consensus and supporting the delegation of authority to experts is a theme running through the development of social regulation.

Many aspects of thought about regulation changed between the 1890s and the 1930s. Two developments we discuss below were especially significant: the emergence of the pluralist language as an alternative to the managerial and the persistence of localism that undergirds the communitarian discourse on regulation.[9] Nevertheless, belief in a discoverable public interest, the core of the managerial language, remained an influential guide in the establishment of federal regulatory agencies during the New Deal. Hanson argues persuasively that the New Deal transformed American political rhetoric by fully developing the notion that the production and consumption of goods and services—not moral claims about the nature of democratic government—were the basis of state legitimacy.[10] Thus government would be judged by its success at fostering production, and citizens would be transformed from participants in the political process to consumers of the outcomes of those processes. The rise of "democratic consumerism" placed great power in the hands of technocratic experts who could manipulate private markets to maximize production but paid scant attention to participation by citizens.[11]

James Landis's thoughts about regulation illustrate these points. He assumed that unregulated private markets could not produce socially just or efficient outcomes. There was much disagreement within the New Deal over regulatory strategies, yet "the single overarching idea that tied the competing philosophies together was the conviction shared by a majority of New Dealers that economic regulation by expert commission could bring about just results."[12] Thus, in expertise would lie the source for discovering policies that would serve the public interest.

In his influential defense of regulation, *The Administrative Process*, Landis argued that only experts can be sufficiently familiar with the issues involved in the regulation of specific industries.[13] What makes this circumstance compatible with American democracy is the argument that such decision makers are more likely than legislators, executives, or the courts to produce policies consistent with the public interest. In the following passage, Landis—with almost religious fervor— links the need for government by experts with the search for policies that serve the public interest:

Government today no longer dares to rely for its administration upon the casual office-seeker. Into its services it now seeks to bring men of professional attainment in various fields and to make that service such that they will envisage governance as a career. The desires of these men to share in the mediation of human claims cannot be denied; their contributions dare not casually be tossed aside. . . . The rise of the administrative process represented the hope that policies to shape such fields could most adequately be developed by men bred to the facts. That hope is still dominant, but its possession bears no threat to our ideal of the "supremacy of law." Instead, it lifts it to new heights where the great judge, like a conductor of a many-tongued symphony, from what would otherwise be discord, makes known, through the voice of many instruments the vision that has been given him of man's destiny upon this earth.[14]

As with Adams, the assumption of a discoverable public interest in regulatory issues allows Landis to avoid the troubling issue of the degree to which democratic values are contradicted by regulatory policy-making within administrative agencies. In a manner characteristic of the managerial discourse, democracy is viewed as a mode of decision making designed to arrive at the public interest: it is a means to an end, rather than an end in itself. If the public interest is better served by alternative decision-making mechanisms, democratic procedures should be abandoned.

As regulatory strategies evolved in the decades between the New Deal and what McCraw calls the "economists' hour" of the 1970s, the public interest continued to hold a central place in regulatory debates. This new hour is based on the belief that economic theory provides a sensible guide to regulatory policy-making. Alfred Kahn's application of marginal cost principles to utility and transportation regulation, for example, was based on the idea that they constitute the most efficient guide to regulation of these industries. *Economic efficiency* was another term for *the public interest;* Kahn championed "the forces seeking to implement economic efficiency for the broad benefit of American society."[15]

Kahn's ideas resonate with both democratic consumerism and Progressive beliefs as he conjoins increased consumption with the public interest. In this view, the appropriate criterion for evaluating the political system is the consumption achievements it produces. Consistent with "the end of ideology" debate that took place during the same period, consumption goals, according to Kahn, should transcend partisan politics and thus are best handled by hardheaded technocrats.[16] What changed in the decades between Adams and Kahn was not the conviction that experts can make regulatory policy in the public interest, but the nature of the expertise thought most likely to produce those desirable policies.[17]

Although much changed in the managerial discourse from the 1890s to the

1970s, a constant was the central assumption of a discoverable public interest that defines regulatory policy as a technical rather than a political problem. In their belief that technical knowledge wielded by experts located within bureaucratic agencies can rationalize regulatory policy, Adams, Landis, and Kahn all reflect the underlying assumptions of the managerial language. In particular, the criterion of economic efficiency is seen not only as a suitable substitute for more democratic methods of decision making but also as more likely to produce policies consistent with the public interest. This conclusion depends upon the tautological argument that the public interest is defined as the achievement of economic efficiency. Both pluralist and communitarian languages challenge the degree to which, given the structural realities of American politics as they identify them, expertise can ever truly be an independent, neutral guide to decision making.

The case of the fourth prophet of regulation, Louis D. Brandeis, illustrates the tensions that exist between economic efficiency and alternative criteria for judging the public interest. Brandeis's experience and success as the "people's lawyer" (representing, among others, small business interests that were coming under increased pressures from large bureaucratic corporations) gave him, in sharp contrast to the other three prophets of regulation, who were hostile to the tugging and hauling of political processes, a "love for the advocacy system."[18] His support for strong antitrust legislation and for the creation of the Federal Trade Commission as well as his opposition to price cutting, used by large firms to the disadvantage of small ones, represents an attempt to protect, through government regulation, values and interests challenged by the emergence of the modern American economy.

Viewed solely from the perspective of economic efficiency, Brandeis's reasoning had serious flaws.[19] However, appreciation of his criticisms requires shifting perspective away from the assumptions of the managerial discourse. That is, underlying much of Brandeis's arguments was a reliance on due process rather than economic efficiency as a criterion for judging regulation. For Brandeis, economic efficiency was not the only end of government intervention in the economy. He worried that the domination of the private sector by large corporations threatened not just the small businessperson, but democratic values as well. Thus, he wrote, "I have considered and do consider, that the proposition that mere bigness can not be an offense against society is false, because I believe that our society, which rests upon democracy, cannot endure under such conditions."[20]

Brandeis's arguments raise issues that define the limits of the managerial perspective because the expertise of bureaucratic managers provides no basis for resolving fundamental value conflicts. To the extent that he was concerned with the political implications of concentrating economic power in large corporations, the economic efficiency of such firms is irrelevant. Further, while economists or other trained experts can define economic efficiency and develop policies designed to further it, when competing noneconomic values are at stake, objectively defined standards for judging policy alternatives do not exist by definition: this is especially so in the new social regulation.

The New Social Regulation and the Limits of the Managerial Language: Risk, Redistribution, and the Public Interest

Although the 1970s saw the economists' hour, it also saw the thrust of much regulatory activity shift from economic to social regulation. New agencies like the Occupational Safety and Health Administration (OSHA) and the Environmental Protection Agency (EPA) were established to deal with "the externalities and impacts of economic activity"[21] on the environment and the workplace. The Consumer Product Safety Commission and a renewed Federal Trade Commission attended to market failures associated with inadequate information and unsafe products offered to consumers. The changing nature of regulatory concerns, evidenced by the rise of what is often called the new social regulation, challenged the validity of managerial concepts of the public interest upon which so much regulatory theory is based.

The ability to distinguish objectively between political issues (appropriately dealt with by electoral institutions) and technical issues (appropriately delegated to experts) is a central assumption of the managerial discourse. It underlies the belief that most regulatory policy is technical in nature and best dealt with by experts. However, when it comes to social regulation, this distinction is far from clear. Indeed, the very act of distinguishing between technical and political issues is open to challenge and becomes, itself, a political issue.

We have seen that, consistent with democratic consumerism, increased consumption has been defined as the public interest and used as the primary criterion for judging regulatory policy. Social regulation, however, is particularly incompatible with the assumptions of democratic consumerism because it involves, in large part, the negative effects of increased production and consumption (for example, environmental degradation or workers' exposure to toxic hazards). Social regulation attempts to remedy the failure of the private market to price adequately the negative externalities of many productive practices: for example, costs of pollution and many types of unsafe products. Moreover, many of these externalities— human life and environmental quality, to name just two—may inherently resist economic evaluation. This failure is crucial because it creates a situation that demands public action. Even government inactivity is a form of action (or nondecision) because it ratifies the failure of markets and imposes economically unjustifiable costs on certain groups in the population (for example, workers at risk, unwary consumers, those who value an unpolluted environment).

Three specific characteristics of the new social regulation make the distinction between technical and political issues problematic, rendering the assumptions of the managerial language quite difficult to defend: scientific uncertainty, risk, and redistributive effects. Environmental, safety, and health regulations often involve levels of scientific uncertainty that render probability estimates of risks quite difficult. Efforts at estimating risks are complicated further by the fact that causes may be separated from effects by decades.[22] Owing to high levels of uncertainty, coupled with the necessity of imposing risk of an unspecifiable probability on some segment of the population, there is no independent, objective standard for judging such decisions. When this uncertainty accompanies a conflict between groups that are fighting to avoid the consequences of regulatory policy, then there can be no legitimate appeal to a public interest discoverable by experts—be they scientists, economists, or lawyers.

Consider, for example, attempts to regulate carcinogens in the workplace.[23] First, scientists are divided over the proper model for explaining carcinogenesis: are one-shot or threshold models more appropriate?[24] Disagreement also exists over the proportion of cancers that can be explained by environmental, genetic, or other causes.[25] Following from this, there is a great deal of uncertainty as to appropriate procedures for estimating the relation between given levels of exposure to a particular substance and actual cases of cancer. Resulting probability estimates can differ by several orders of magnitude, depending upon assumptions about the relation between tests on laboratory animals and human disease or the appropriate implications to be deduced from limited epidemiological studies.[26] Regulatory agencies charged with addressing such issues differ significantly in the assumptions they make and the estimates of risks and costs they produce.[27]

Uncertainty in social regulation complicates an already problematic feature of all regulatory policies: their redistributive impact. Because they attempt to alter the outcome of market-determined distributions of income, such policies always redistribute income from one group to another.[28] Those groups forced to bear the costs of regulatory policy are likely to object, even when these costs are outweighed by the benefits such policies might bestow on other groups.

The potential of redistributive impacts to give rise to political conflict is precisely what made the notion of a discoverable public interest, implicit in the managerial discourse, so attractive to regulatory reformers. Adams and Landis, for example, argued that regulation would, over the long run, produce benefits even for those groups forced to bear costs in the short run (for example, railroad owners or stock exchanges). More recent regulatory reformers, Kahn, for one, imply that the public interest would be served if the costs imposed on certain groups were outweighed by the benefits reaped by other groups. The Reagan administration's Executive Order 12,291, for example, charged the Office of Management and Budget with applying such a standard to most regulations proposed by executive agencies. Such a standard, applied by administrators, defines regulatory policy as simply applying market standards of efficiency when markets cannot produce such results on their own.

Although such standards of regulatory policy-making may be plausible for certain types of economic regulation, they are constantly challenged in the new social regulation. Environmental, consumer, health, and safety regulation all entail slippery quality-of-life questions that make it difficult, even in theory, to capture the value of the resources at issue.[29] The issues in airline or utility regulation, for example, involve questions concerning who will bear the monetary costs of monopoly, what are the economies of scale, and how they can be fairly distributed among consumers and producers. In contrast, social regulation addresses such questions as who will risk getting cancer in twenty-five years, what are the effects on groundwater of contamination with various hazardous substances, who will live near nuclear power plants, and what value do we place on a healthy environment.

In spite of its inability to capture the dynamics of the new social regulation, the managerial language continues to be used by many policymakers and retains its power in policy disputes. Indeed, those who adopt the managerial discourse, by appealing to scientific or technical expertise, often appear to stand above the petty squabbles of partisan debates couched in pluralist or communitarian languages. So, for example, in the case of the EPA, an "accent on scientific and technical analysis tended to endow [the agency] with a certain credibility not available to other agencies cast in the adversarial culture of the legal profession alone."[30]

Policymakers also appeal to Adams's notion that conflicts can be avoided and consensus achieved if the public is "properly" educated before it is asked to participate. Reflecting this view, William Ruckleshaus, the first head of EPA, stated, "Newer, more complex issues at EPA demand public education. . . . Cost-free public involvement is a recipe for public agitation. . . . The agency must educate the people to think in terms of efficient risk management rather than the elimination of risk."[31]

The persistence of the managerial motif notwithstanding, the political issues raised by social regulation lead to the appearance of other languages in debates. Pluralist and communitarian discourses focus, in ways that the managerial language does not, precisely on these political questions.

The Pluralist Language

While regulators themselves emphasize technical issues and tend to understand regulation in the terms set by the managerial discourse, this is not the only language available for debating regulatory policies. Especially since the New Deal, some policymakers (particularly elected officials) and many scholars use what we call a pluralist language to emphasize the political dilemmas of this area of public policy. Pluralist language draws heavily upon the arguments of mainstream academic theorizing on the relation between private markets and democracy and therefore is familiar to political scientists.[32] Unlike the managerial language, pluralist discourse assumes that conflicting interests are the essence of politics and, in a democracy, cannot be resolved by appeal to an overriding public interest discoverable by experts. Instead, the public interest is served by creating an open political process that allows contending organized interests equal opportunity to influence public policy. Adopting the logic of market relations and its political analog in the group competition of pluralist politics, this perspective assumes equality of opportunity, political equality, and freedom.

These assumptions undergird pluralist discussions of social regulation. It is assumed that if the political process is open, then the outcomes of regulatory policy will be fair. Implicit in such a view is the notion that liberal democratic political institutions, if structured correctly, are capable of altering the outcomes of private market processes and remedying market failures. Absent a belief in a public interest discoverable a priori, good regulatory policy emerges not from the delegation of authority to neutral experts, but rather from policy-making procedures that allow open access to all affected interests. In this way, regulatory policy balances the vectors of political pressure brought to bear by organized interest groups.

In regulatory policy, once legislation is passed, group pressure focuses on the bureaucratically organized agencies charged with implementation. Because of this, both managerial and pluralist approaches to regulation remain in uneasy coexistence as similar concepts and solutions take on different meanings within each discourse.[33] Policy confusion often results from the failure to recognize the contested nature of these concepts. From the managerial perspective, the vague injunction to regulate in the public interest, characteristic of regulatory legislation prior to the 1970s, meant that decisions should be made by expert regulators. To pluralists, this same injunction means that legislators or experts are unable to determine the public interest and the only solution is to design a decision-making process that will allow the balance of group interests to define the public interest. The failure to resolve differences in a term's meaning in different languages is an enduring theme in social regulatory policy.

Whereas debate within the managerial language revolves around the most appropriate source of expertise necessary to discover the public interest, debate within the pluralist language turns on the best way to structure the policy process so as to balance contending interests. Most contemporary scholarly debate between conservatives and liberals over social regulation turns on this issue and is a reaction to the perceived inability of existing regulatory policy to achieve such a balance. These two camps differ over whether due process must be in the form of democratic participation in the political realm (the liberal position) or whether citizens' interests can be served by their participation in private markets (the conservative position). Nevertheless, for both liberals and conservatives, debate remains firmly within the pluralist perspective: the public interest is defined not as a specific outcome, but rather "procedurally," as a set of fair and open allocational decision-making processes.

Conservative theorists argue that capitalism provides the necessary preconditions for democracy.[34] The ability of private markets to transform the individual pursuit of self-interest into an efficient social allocation of resources allows the political freedom of minimal liberal democratic government. Freedom, for these theorists, is the ability to maximize one's self-interest, and markets provide an arena of maximum freedom because they allow the tailoring of consumption to individual desires. The ways in which such desires may be manipulated or ignored—thus limiting their expression—by such institutions as advertising or the mass media are concerns little addressed by conservative theorists and typically lie outside the pluralist discourse.

By ignoring the possibility of coercion in the private sector—one of Brandeis's concerns—conservatives are able to contrast sharply market allocations of goods and services with those of the public sector.[35] They argue that because government policies are imposed on all citizens, political decisions cannot be accurately tailored to individual preferences as can market activities. The former, therefore, must involve some element of coercion that is absent from the latter. Resting their arguments on the superiority of market decisions to political decisions, these theorists maintain that individuals can be politically free only because of the possibilities for economic freedom opened up by the market.

Regulation, because it responds to the actual or perceived failure of market mechanisms to allocate resources efficiently, poses a challenge to such theorists. When markets fail—when they do not aggregate individual self-interest into a social good—the justification for allowing them to allocate society's resources is undercut. Market failures demand political remedies because inaction on the part of government perpetuates the inefficient allocation of resources and can also lead to abuses of property rights and the individual liberty upon which such rights rest.[36]

If capitalism is compatible with democracy, indeed if it creates the preconditions for individual liberty, then it must be possible for democratic political systems to regulate private markets when they do fail. Conservative theorists reluctantly concede the need for regulation in some instances. They argue, however, that it is used much too frequently and is often based upon inappropriate criteria developed within a faulty policy-making structure. This line of reasoning and its attack on existing regulatory policies underlies the Reagan and Bush administrations' program for deregulation. Here again conservatives believe that individual freedom is best served through maximum reliance on economic markets and minimum resort to more easily distorted political markets.[37] Nevertheless, the validity of this approach rests on the belief that properly designed political institutions are capable of remedying or preventing the more serious consequences of market failures. Part of this proper design involves the creation of effective limits on the intervention of the state into private markets.

Conservatives generally argue that such limits ought to be defined by the market criterion of economic efficiency: the costs of regulation should not exceed the benefits. However, insuring the application of this criterion creates a paradox. Because regulation arises in connection with market failure, in those instances normal market processes cannot be relied upon. Further, the unhindered operation of pluralist politics is subject to systematic distortions that militate against policy outcomes that are consistent with economic efficiency. First, there are significant structural barriers to the political organization and representation of large "latent" groups.[38] The costs and benefits that fall on such groups will be subordinated in the political process to the interests of smaller, privileged groups. Second, market failures themselves may create political conditions that reinforce this distortion. To the degree that economic resources translate into political power, those who benefit from market failures may be able to use those benefits to protect their position politically. In the absence of any independent standard for judging the public interest, regulatory policy-making is likely to become a battleground for the organized groups that have a stake in the redistributive impacts of such decisions. Regulation, in George Stigler's apt characterization, becomes a good that can be purchased from government for private benefit.[39] Thus, market failure may establish the preconditions for "government failure."[40]

The pluralist language alone does not provide a solution to this problem for conservatives. As we have seen above, conservatives often implicitly or explicitly assume that experts schooled in marginal utility economics and insulated from pluralist politics within bureaucratic agencies are capable of imposing the criterion of economic efficiency and thereby limiting government regulation of the economy. Thus, while this approach to regulation begins in the pluralist perspective, it moves quickly into the purview of the managerial language, which assumes the inevitable expansion of state power so feared and criticized by conservative theorists. That is, the unregulated dynamics of pluralist politics and private market forces leads away from a reliance on the criteria for policy-making espoused by conservatives. The solution is to use the powers of the state to enforce these particular standards of regulatory policy. Ironically, then, the expansion of state power and authority becomes the only way to prevent the increased intervention of the state in the economy.

Richard Harris's and Sidney Milkis's summary of the Reagan administration's attempt to enforce cost/benefit standards in order to reduce regulatory burdens captures the dilemma facing conservatives: "The Reagan pursuit of a 'new social contract' may, in the end, undermine the prospects of enduring change. It has, most obviously, put the advocates of 'real economics' [that is, marginal cost economics] in the uncomfortable position of carrying out a program to reduce the burden of government regulation through what amounts to unprecedented administrative aggrandizement. Regulatory relief, therefore, has led not to institutional reform that imposes restraint on government action, but to an important confirmation of a substantial government presence in overseeing social and economic activity."[41]

Liberal criticism of existing regulatory institutions uses pluralist discourse differently. Disagreeing with conservatives, liberal critics argue that the relation between capitalism and democracy is historical rather than logical. Both systems, they maintain, grew out of reactions to a feudal order that did not allow for individual freedom—either political or economic. There are, however, inherent tensions between the two systems. In general, the emphasis of market economies on the individual pursuit of self-interest is accused of weakening the ability of polities to act for collective purposes.[42] In particular, the economic inequality resulting from market competition leads to forms of political inequality that are incompatible with democratic theory. Marxists are the most well known advocates of such a position, but mainstream social scientists have also used the pluralist perspective to question the compatibility of democracy and capitalism.[43]

Charles Lindblom provides an excellent example of this position when he suggests that the private market creates a prison from which policymakers in a democratic political system cannot escape. Regulation is troublesome for democratic political leaders; interference in the market is often "rewarded" with inflation, unemployment, slow growth, or all three, which ultimately lead to defeat for such leaders at the polls. In particular, Lindblom emphasizes the significance of the emergence of large-scale bureaucratic corporations. Echoing the concerns of Brandeis, he argues that these institutions violate the assumptions of both economic theories of the competitive market and political theories of democracy; they produce neither the liberty nor efficiency of the market while violating the norms of democratic political processes. Lindblom concludes that the relation between democracy and capitalism in its modern form is, at best, tenuous and uneasy.[44]

With respect to regulation, liberal critics hold that the economic and political power of corporations makes it difficult for political leaders to impose regulatory costs on these institutions. If regulation deals with market failures and if the dominant economic institutions of modern capitalism make it impossible for democratic political institutions to address such failures, then modern capitalism may be incompatible with democratic politics. For such theorists, the central problem of regulatory policy is not limiting the powers of government to intervene in the market, but rather limiting the ways in which unequal distributions of economic and political power distort the actions of government, thereby making true due process impossible. Liberal reformers typically advocate political programs, which they assume can be adopted by pluralist politics rightly structured, limiting the impact of economic power on the political process.[45] Typical of pluralist reforms, such programs involve the inclusion of a wider array of interest groups more capable of representing the full range of interests affected by regulation.

As with the managerial language, the pluralist perspective alone is inadequate for understanding social regulation. The nature of the issues involved (in particular, the scientific and technical complexity of comprehending the risks addressed by social regulation) creates severe barriers to participation because access to such information is a prerequisite to understanding one's interests. Further, there are structural problems posed by the redistributive implications of such regulation. Because the distribution and redistribution of risks is often highly correlated with economic inequality, relying on due process as conceived within the pluralist language can result in regulatory decisions that are severely skewed and patently inequitable. These impediments make problematic a core pluralist assumption that the absence of participation indicates consensus. The lack of participation by many unorganized interests is a departure point for the communitarian discourse, which challenges both managerial and pluralist understandings of social regulation.

The Communitarian Language

Not surprisingly, the pluralist and managerial discourses dominate debate about social regulation: they are used by policymakers, scholars, and citizens to discuss and debate most political issues in the United States. In this debate, they engage each other and coexist uneasily. Sometimes there is genuine negotiation across languages with the development of common understandings; at other times, the same terms mean very different things in the two languages and are subject to political manipulation.[46] Yet these two languages, either alone or combined, cannot capture all the issues raised by social regulation or by the way in which these issues are actually debated, especially at the state and local levels. When regulatory decisions are disputed, challengers often use a vocabulary that rejects the basic assumptions of both pluralist and managerial discourses. This language of challenge was difficult for us to name and define because it rarely surfaces in elite debates or in the academic literature, and it does not engage the other two languages in the design of regulatory institutions. Neglect of what we call the communitarian language has deep roots in American political ideology.

Shlomo Avineri and Avner De-Shalit describe the communitarian challenge to the underlying assumptions and normative visions of both the pluralist and managerial discourses:

Methodological[ly], . . . communitarians argu[e] that the premises of individualism such as the rational individual who chooses freely [which underlies both managerial and pluralist vocabularies] are wrong or false, and that the only way to understand human behavior is to refer to individuals in their social, cultural, and historical contexts. That is to say, in order to discuss individuals one must look first at their communities and their communal relationships. . . . normative[ly], communitarians assert . . . that the premises of individualism give rise to morally unsatisfactory consequences. Among them are the impossibility of achieving a genuine community, the neglect of some ideas of the good life that should be sustained by the state, or others that should be dismissed, or—as some communitarians argue—an unjust distribution of goods. In other words, the community is a good that people should seek for several reasons and should not be dismissed.[47]

The yearning for community and the critique of liberal democracy such desire implies have deep roots in American political history. Jennifer Nedelsky, focusing on the politics of James Wilson, a Founding Father from Virginia, argues that a communitarian voice was raised at the Constitutional Convention in opposition to the more limited democratic dialogue of the other Founders. She contrasts Wilson's emphasis on civic participation as the basis of government's constitutional structure with James Madison's (and the Federalists') emphasis upon the protection of private property from the tyranny of the majority.[48] One can easily discern the struggle between the communitarian language of politics, on the one hand, and the managerial and pluralist languages, on the other, in the following passage dealing with what Nedelsky calls "the failure of public liberty":

We are left then with a kind of self-fulfilling Federalist vision of politics: They were afraid the people would use their political power to pursue their private interests at the expense of the rights of others and of the public good; they created a system in which the only possible relation between citizen and representative was that of advocate and arbitrator of interests. The institutions did not foster the ongoing engagement with public affairs that could have transformed raw private interest into considered political opinion. That was the Wilsonian solution to the democratic threat that his fellow Federalists could not comprehend. They were right that unmediated private interest backed by political power was dangerous, but they focused on containing rather than transforming the threat. Citizens' limited perspectives were not to be enlarged, but contained or refined by representatives capable of an approach to politics which "the people" were not.[49]

Tocqueville too noted the conflict between liberalism and communitarian visions when he remarked on the tensions in America between the centrifugal forces of individualism reinforced by the emergence of capitalism and the centripetal attachments of religious and family ties rooted in community. By the late twentieth century, acceleration of geographic mobility driven by the revolutionary forces of capitalism and the market, the emergence of a mass media and widely shared popular culture, and other social forces tend to obscure the degree to which the affective ties of community still resonate for many Americans.[50] In *Habits of the Heart*, Robert Bellah and his colleagues revisited Tocqueville's observations to document the American yearning for the lost values springing from attachment to a political and social community. Indeed, they note that a characteristic way some of their subjects try to recover this lost sense of community is by becoming politically active in local affairs.

The political language that grows out of the communitarian critique of limited citizenship is outlined in *The Democratic Wish*, James Morone's study of the long-running tension between attempts to perfect a more participatory form of democratic governance in the United States and the dramatic growth of public bureaucracy.[51] Like Nedelsky, Morone argues that American democratic thought has historically been torn between, on the one hand, a fear of public power and the threat it holds for freedom and, on the other hand, a desire for communal democracy. In reaction to the centralization of power in Britain that was seen as the cause of the American Revolution, there has always been a belief that American democracy must avoid large, strong governmental or private institutions. Such institutions pose a grievous threat to individual freedom (political and economic) and need to be viewed with great suspicion.[52] In this view, the delegation of authority to governmental institutions causes them to grow uncontrollably and is incompatible with true political democracy.

Instead, as Wilson argued, the "democratic wish" holds that it is possible to sustain an enlightened citizenry capable of ruling directly through communal forms of democracy. The creation of citizens capable of overcoming narrow self-interest would allow doing away with illegitimate political institutions that are removed from popular control and would return government to its true source of legitimacy: the people. Thus, the overriding public interest will be discovered by this enlightened citizenry: apparent conflicts of interest among the people will disappear in the processes of communal self-government. This utopian call for returning government to the people is a thread running throughout American political history, surfacing during periods of political reform and providing a potent language for challenging existing political and economic relations.

Our identification of the communitarian language as a distinctive and important component of debate over social regulation owes much to our focus on the impact of these policies on geographically defined communities. We believe that the reaction of local communities to the policies created at the federal and state level is an important component of the development of regulatory policies in the

United States but is often overlooked in the scholarly focus on the national level.[53] Given the communitarian language's emphasis on returning government to the citizenry and its hostility toward the centralization of political authority, it has been attractive to defenders of local cultural and political values in their attack on the legitimacy of a strong national state.

The communitarian language, and the democratic wish embedded within it, offer an alternative to managerial and pluralist languages. As we have seen, the managerial language assumes a public interest discoverable by neutral experts, while the pluralist discourse rejects such a priori specification and sees the public interest emerging from competing forces in an open political process. In contrast, communitarian thought suggests that conflicts of interest among the polity are illusory, and a common public interest can be discovered if an enlightened citizenry governs directly in its own behalf. The delegation of authority to experts or the reliance on competing pressure groups distances government from the people and stands in the way of perfecting democracy and citizenship.[54]

The communitarian language is central to the recurring dynamic of political reform in the United States, and Morone traces its influence in the Jacksonian, Progressive, and New Deal periods. Throughout American history, economic and social changes have produced new interests not taken into account by existing political and economic arrangements. Political conflicts develop as these arrangements are challenged. A potent tool available to those who challenge the status quo is the logic of the democratic wish, with its deep resonance in American political thought. Its appeal reflects the degree to which existing political and economic institutions are always vulnerable to the charge that they are unrepresentative and illegitimate because distant from the citizenry.

Reformers, maintaining that the perfecting of democracy requires widespread participation and public policies directly responsive to the people, contend that government and political power must be returned directly to the citizenry. Yet Morone argues that the democratic wish has the ironic result of accelerating the very bureaucratization of the state attacked by reformers. This is so because reformers, beyond calling for an enlightened citizenry capable of ruling on its own, seldom detail plans for carrying out this dramatic reformation of the American political system. That is, the communitarian language fails to specify either how an enlightened citizenry might be created and maintained or the nature of the specific institutions that would exercise power. Appeals to communitarian ideals can mobilize popular support for political change, but reforms do not establish the prerequisites for a transformed polity. Rather, falling back into the terrain of pluralist and managerial discourses, reforms create new access routes for organized political interests to influence the bureaucracies charged with making and implementing public policy.

Reform, then, fails to create the communal institutions necessary to bring government closer to the people; instead, it leads to newly created bureaucratic organizations and interest groups that operate on behalf of reformers' conception of the citizenry. The result is a new round of institutional growth that makes government even more vulnerable to criticisms based on the democratic wish.

Regulatory reform during the 1970s, which led to the rise of the new social regulation, is an example of the democratic wish in action. Reformers employed the communitarian language to articulate their criticisms of existing regulatory policies and institutions. However, the inability of the communitarian language to address the task of creating a citizenry able to take an active role in the formation of regulatory policy led to new regulatory institutions that relied upon pluralist and/or managerial solutions. This, in turn, gave rise to the charge of grassroots groups that social regulation, like older forms of regulation, was insensitive to their interests and barred them from participating.

During the 1970s, the emergence of "the public interest lobby" helped place the issues addressed by the new social regulation on the political agenda. The lobby's success reflected reformers' ability to exploit the power of the communitarian language to tap widespread longing for the establishment of an active, enlightened citizenry. Echoing the democratic wish, reformers at the head of groups pressing for consumer protection, environmental protection, and other issues central to the new social regulation challenged the institutions, public and private, that dominated policy-making, claiming that they had become too large and removed from popular control to be legitimate or to produce policies consistent with the public interest. On the one hand, they criticized existing regulatory processes for being dominated by powerful economic interest groups that prevented the broader interests of the people from being represented. On the other hand, especially in the area of environmental protection, reformers challenged the ability of experts, especially those employed by government and business, to capture the values involved in protecting the environment.

Events like the first Earth Day in 1970 marked the emergence of environmentalism as a powerful social movement reflecting a more general questioning of the worth of technological progress and economic growth, central values underlying both the managerial and pluralist discourses. Environmental groups rejected the ability of experts to capture the values at stake in vital environmental issues or the possibility of balancing such values against other competing economic considerations. Instead, articulating the democratic wish, they called for participatory democracy, arguing that only by including the people directly in the regulatory process could environmental values be preserved. Further, they restated the communitarian discourse's basic distrust of representative institutions, large public and private bureaucracies, and the delegation of authority to experts.

This discourse has often been portrayed by conservatives and other defenders of the status quo as being rooted in Marxist or socialist ideologies that have little relevance or resonance in the American context. We believe it is more accurate to see it as a reemergence of the communitarian language that is an enduring part of American political rhetoric. In essence, a suspicion of large institutions, both political and economic, along with a yearning for participatory democracy have always been central components of American political ideology. It is this continuity with such traditional themes that accounts for the popular appeal of reformers and their ability to force accommodations from the political system in the shape of redesigned social regulatory institutions.[55]

So, while the communitarian language gave the claims of the public interest lobby movement great power and popular appeal, it also meant that the new social regulation would reflect the basic limits of that discourse. As we detail in the next section, public interest groups were successful at forging a role for themselves, as *organized interests,* in the decision-making processes of the new and reformed bureaucratic agencies that would shape social regulation. However, this was far short of the call originally articulated by reformers for direct participation by the citizenry in matters that directly affected their lives. Reflecting the repeating dynamic of reform inspired by the democratic wish, the new social regulation is marked not by an increase in citizen participation, but rather by the growth of bureaucracy and the increasing influence of large, organized interest groups. Like the pluralist and managerial discourses, the communitarian language is limited in that it can pose such challenges but does not provide the tools for meeting them.

The inability to chart a specific course of action is inherent in the nature of the appeals for popular mobilization articulated within the communitarian discourse. In defining the people as distinct from the practices of political and economic institutions, the communitarian language implicitly assumes that the people have a single set of interests that are ignored by those institutions. In environmental and consumer protection this was a persuasive rhetorical device, because all citizens presumably share an interest in reducing pollution and eliminating unsafe and deceptive practices. However, as regulatory legislation is written and, more important, as specific rules are developed and implemented by government agencies, conflicts of interest among different segments of the citizenry rapidly develop over how to redistribute the costs and benefits of social regulation. If the communitarian call for participatory democracy is to be taken seriously, mechanisms must be specified for developing a citizenry capable of participating in the development of a consensus over such trade-off decisions. Yet the rhetoric of regulatory reform ignored these sorts of issues. While claiming to represent the overriding interest of all citizens (as consumers or victims of pollution), reformers actually opted to enter the policy process via more traditional interest group politics.

In sum, the communitarian language is effective in challenging the legitimacy of existing institutions because of its deep roots in American political thought. Yet, because it does not suggest concrete solutions, the design of institutional responses to the communitarian challenge falls back upon the logic of pluralist or managerial discourses. Ironically, this makes the decisions produced by institutions vulnerable to continuing rounds of attack by grassroots groups speaking the communitarian language.

The Limits of Managerial, Pluralist, and Communitarian Discourses: Interest Group Liberalism and the Declining Legitimacy of Social Regulation

Although the social regulation of the 1970s was inspired by the communitarian challenge of the public lobby, institutional responses resulted in a new regulatory regime that altered, but did not transform, the interest group liberalism that continues to dominate policy-making.[56] Reforms extended participation in policy-making to interest groups working on behalf of the public in their roles as consumers or environmentalists but failed to address fundamental questions about citizenship and participation raised in the communitarian rhetoric of reformers.

Understanding the influence of managerial and pluralist rhetoric on the institutional reforms that constitute the new social regulation requires understanding the structure of interest group liberalism and its continued sway in American policy-making. Theodore Lowi, in *The End of Liberalism*, argues that a fundamental political problem in the United States is that, as government has expanded in the postwar era, legislative bodies—and elected leaders in general—have delegated increasing amounts of policy-making authority to administrative agencies.[57] Unwilling to make difficult, specific political decisions, elected leaders pass legislation that shifts policy-making from legislatures and executives to bureaucracies. Consequently, policy-making is dominated by interest groups with the political resources to gain access to administrative agencies. Lowi calls this system "interest group liberalism": "interest group," because only those segments of society represented by organized groups can influence the policy process; "liberal," because the resulting fragmentation and specialization of policy-making leads to the expansion of government.

Especially relevant here is Lowi's argument that the existing languages available to debate public policies in the United States are inadequate. While "interest group liberalism" represents the new public philosophy, the rhetoric of politicians and the public is still organized around the now inappropriate liberal-conservative debate over the very existence of positive government. This debate fails to capture the actual dynamics of political power and struggle and thus creates a crisis of public authority in the United States.[58]

Reflecting the dominance of both managerial and pluralist approaches among elites, a variant of interest group liberalism remains the characteristic style of policy-making within social regulatory agencies. James Q. Wilson developed a useful and widely used framework for analyzing the dynamics of interest group liberalism within the regulatory arena.[59] He argues that an important characteristic of social regulation is that it attempts to impose concentrated costs on certain groups (for example, owners of industries producing pollution) in order to benefit a diffuse and unorganized latent group (for example, all individuals affected by pollution produced by given industries). Passage of such legislation requires the construction of majorities capable of overcoming the inevitable resistance of the concentrated, and therefore usually organized, groups who stand to lose from such regulation. The passage of such legislation is something of a political puzzle: "Since the incentive to organize is strong for opponents of the policy but weak for the beneficiaries, and since the political system provides many points at which opposition can be registered, it may seem astonishing that regulatory legislation of this sort is ever passed."[60]

This puzzle has two solutions. Both help explain how the use and limitations of the three languages used to debate social regulation reproduce interest group liberalism. First, Wilson suggests that such legislation is usually passed through the work of political entrepreneurs (for example, Ralph Nader and other leaders of the public interest lobby movement) who, acting as "the vicarious representatives of groups not part of the legislative process," take advantage of crisis or scandal to mobilize support for regulations. As we have seen, these entrepreneurs often employ the communitarian language, condemning political and economic institutions and calling for a return of power to citizens. Such mobilization and the power of the democratic wish make it difficult for politicians to oppose passage of legislation. Indeed, many politicians have seized upon such issues to build themselves constituencies for election to public office.

Second, according to Wilson, entrepreneurial politics leads to extreme regulatory legislation because it depends upon current outrage over a crisis to build majorities. However, Wilson fails to analyze fully the limitations of the language that gives rise to such extremism or the dynamics of interest group liberalism within which social regulation is implemented. One of the differences between the new social regulation and older forms of economic regulation is that Congress is less likely in the first case to give agencies vague charges to act in the public interest. Rather, reacting to crisis and to the rhetoric of entrepreneurs, legislation provides specific guides for policy and calls upon agencies to create a regulatory framework within a specified amount of time. Nevertheless, the implementation of this framework—the specific substances or products to be regulated, safety and risk levels, calculations of costs and benefits—are left to the agency. That is, what is most political about social regulation is to be decided by administrative agencies.

The Delaney Clause of the Food, Drug, and Cosmetic Act and the National Ambient Air Quality Standards of the Clean Air Act are cited as prime examples of social regulations that reflect the extremism discussed by Wilson. These acts prohibit any balancing of costs and benefits by establishing "goal[s] to be attained regardless of the costs imposed or the effects that may be imparted to other objectives."[61] OSHA's toxic substances regulation, Mine Safety Toxic Materials regulation, the Atomic Energy Act, Clean Water Standards, and Noise Emission Standards all come close to this sort of absolute goal setting. In practice, such standards are impossible to enforce (they imply that all national income should be directed toward pursuing these absolute goals—an obvious impossibility) and "mock serious political discourse."[62] In addition, these impossible standards, because they provide little actual guidance, place greater power in the hands of administrative policymakers and their constituents.[63]

Usually, this delegation of power to administrative agencies is justified by the managerial assumption that the delegated issues are technical. That is, legislation assumes the existence of technical bodies of knowledge that will allow regulators to determine such things as safe exposure levels to specific substances; the appropriate redistribution of risk across various populations; the appropriate trade-off between safety and economic efficiency; and so on. However, since the data addressing these issues do not exist or are highly controversial, agency approximations of the public interest are problematic, and the outcome of these decisions becomes essentially political.

Increases in the autonomy of agencies also result from gaps in the communitarian rhetoric used to pass such measures. Whereas this language contains a powerful critique of existing institutions, it is much less clear on how serious public debate over such issues might be established. Regulations are passed as a result not of ongoing public discourse but of the periodic outrage of the public in response to specific scandals or disasters (recall the Elizabeth chemical explosion) used by policy entrepreneurs. The outcome is extreme and unenforceable regulations that often lead to dissatisfaction with the specific policy outcomes when these standards are, inevitably, compromised in the interest group politics of regulatory agencies.

In the absence of a public language capable of seriously addressing the trade-offs involved in social regulation, no matter how legislatures and regulatory agencies try to redefine such issues in a technical manner, political conflict is simply shifted from the legislative to the bureaucratic realm. This shift of political struggle from one arena to another mitigates the degree to which social regulation (extreme or not) passed by entrepreneurial-style politics can actually overcome the resistance of the well-organized interests upon whom costs are to be imposed. The failure to overcome such resistance explains the continuing dissatisfaction of newly formed grassroots groups with the compromises produced by the participation of public lobbies in the tugging and hauling of bureaucratic politics.

Further, because the communitarian language provides little guidance for creating the institutions necessary to involve citizens continuously in social regulation, the groups influencing regulatory agencies are unlikely to be the same groups that mobilized to pass legislation. Although a crisis or scandal allows political entrepreneurs to mobilize public support for regulations designed to bestow benefits on large latent groups, such organized support is likely to be transitory: these groups are unlikely to remain mobilized for very long. On the other hand, groups who are concentrated and have substantial and specific stakes in regulation—often, the industry to be regulated—have every incentive to remain organized and maintain continuous pressure on the regulatory process.

The incorporation of the public interest lobby within slightly modified regulatory institutions still dominated by pluralist and managerial approaches gave rise to criticisms, from both the left and right, that turned the communitarian language against reformers themselves. First, conservatives in the Reagan administration advocating regulatory relief claimed that the activities of environmental and consumer groups represented only a narrow, special interest (as do all interest groups), challenging the communitarian claim that such groups represent the interests of the people. Further, advocates of regulatory relief could claim that decision making should be shifted to state and local governments, which were closer to the people, if the goal was to represent the interests of citizens.

Second, as the costs of regulation were imposed on geographically bounded communities, newer citizen groups in these communities questioned whether public lobby groups actually represented their interests. As mainstream environmental and consumer groups were incorporated into the regulatory process, they became vulnerable to the claims of grassroots groups using the communitarian language that they were part of unrepresentative and illegitimate political and economic institutions unresponsive to the interests of the people.

At the EPA, for example, as the public interest lobby became a routine participant in rule making, the development of dispute resolution procedures reduced dramatically the conflict between environmental and business interest groups. However, "[such procedures] breed the sort of decision making that continues to insulate policy networks from the sort of full-scale political dialogue about major issues that might be necessary to replenish the lost sense of citizenship in American politics."[64]

Although the communitarian rhetoric used by the public interest lobby movement is effective at garnering support for social regulation, it does not address the structural impediments to the continuous mobilization of such interests. Indeed, by assuming that the public has a uniform interest and by failing to specify how this common interest can be continuously articulated or mobilized, the rhetoric and structural impediments tend to reinforce each other. It is just this problem that accounts for the split between grassroots and established, statewide environmental groups in the states and localities we studied.

Further, federal regulatory policies supported by national environmental groups often assume a common interest exists among all citizens inspired by the communitarian language. But these policies founder at the state and local levels precisely because that is where conflicts of interest between communities—and actual disputes about which parts of the people will bear the costs of a regulatory policy—actually emerge. Yet the language used to mobilize citizens to support such policies fails to specify how to create democratic institutions necessary for working these issues out in participatory fashion. This aspect of social regulation and the limits of the languages used to debate institutional reform are apparent only if we look beyond the national level to the ways these policies affect geographically bounded communities.

Conclusion

The use of the communitarian language to mobilize support for social regulation followed by the reassertion of interest group liberalism results both in an unwillingness by groups at the state and local levels to accept the adverse consequences of the policy process and in government's inability to implement such policies. While the imposition of losses is difficult under any circumstances—nobody likes to lose—it is especially problematic when those who lose believe that they have not been represented. A foundation of democracy is the willingness to compromise and accept disagreeable decisions because of the legitimacy of the process that produced such policies. This willingness, however, is undermined by the inability of existing political institutions to represent adequately all the interests articulated in the debate over social regulatory issues. Further, the languages used in this debate fail to specify how institutions could be modified or how new institutions might be constructed.

Standing alone, none of the three languages is adequate for capturing the dynamics of this contentious policy arena, let alone for resolving disputes raised within it. In the next chapter we develop a more synthetic approach to social regulatory policy guided by what we call a dialogic model. Consistent with our focus on the languages used to discuss public policies, this model views social regulation as an ongoing public dialogue among citizens. Our model assumes that politically relevant truth in a democratic society can emerge only from an open engagement among these languages. Yet, actual policy debates usually exclude certain languages or ignore their underlying and competing assumptions. At a basic level, social regulation involves conflict over the appropriate methods for establishing what will count as truth in the policy process, the nature of political authority, and the very possibility of democracy in a modern society. Each of the alternative languages used by scholars and participants alike reaches different conclusions about these questions. Until these fundamental differences are fully recognized and squarely addressed, the underlying problems in social regulation will remain unresolved. Exploring these issues is the first task of chapter 3. Second, we use the dialogic model to explore the shape of policy-making institutions that might better resolve the problems of social regulation. There, at a more speculative level, we define the sorts of political reforms required to allow a more integrative dialogue to develop.

3 / A Dialogic Model of Social Regulation

Objectivism, Relativism, and Beyond

In chapter 2 we discussed the problems that develop when managerial, pluralist, or communitarian languages are used by contending parties in an area of public policy, like social regulation, involving both contested technical and scientific knowledge as well as conflicts over fundamental political values. In effect, public discourse and public policy remain trapped by the inability of any single language to capture the dynamics of social regulation. First, the managerial language is unable to legitimate its assumption that regulatory issues involve no fundamental value conflicts and are reducible to a technical search for an objectively best policy. Second, the pluralist language fails to deal with either the core of technical and scientific facts that must be incorporated into policy or the unequal resources available for mobilization by affected interests, both of which tend to be neglected in a focus on the balancing of organized interests. Third, the communitarian language specifies neither how its vision of an engaged citizenry will be achieved, nor how a focus on local action can be integrated into the broader American political economy. Moving beyond these weaknesses involves understanding that the limitations of these languages come not simply from the way they articulate specific public policy concerns, but from their more basic and competing definitions of truth, knowledge, and politics. In short, these limitations "go all the way down" and cannot be remedied within the vocabulary of any specific language. At stake is nothing less than competing definitions of what constitutes political truth and the means for evaluating competing truth claims in the political process.

In fact, these contrasting assumptions about truth and knowledge reflect long-standing philosophical debates between objectivist and relativist models of knowledge. One reason policy debates prove intractable is that they raise, but do not explicitly acknowledge, these epistemological issues. The foregrounding of such issues allows one to employ the insights of philosophers and political theorists who point the way toward a new understanding of the importance of dialogue and democratic participation as methods for transcending the battle between objectivism and relativism and the competing political languages flowing from them. Ironically, the communitarian language, the most neglected of the three policy discourses, comes closest to embracing a definition of politically relevant truth upon which we can construct a model of policy-making that transcends debates between objectivist and relativist theories of knowledge.

Our own approach, which we call a dialogic model of social regulation, starts from the assumption that it is impossible for scholars or participants to capture the dynamics of this arena of policy-making within a single discourse. Because this is so, we go on to argue, the challenge for social regulation in a democratic society is the creation of policy-making institutions capable of containing multiple perspectives and bringing into engagement the competing languages used by the various interests involved in policy struggles when they affect geographically bounded communities.

In coming to these conclusions, we were influenced by developments in a wide variety of fields that have focused scholarly attention on the importance of discourse and how it can or cannot be structured in a democratic society. Particularly relevant for our inquiry is the work of postmodernists, who challenge the modernist belief that there is or can be a single perspective or language that can be used to describe and define complex social and political realities. While the postmodern position is associated with the work of such European scholars as Jacques Derrida, Jean-François Lyotard, and Michel Foucault and such Americans as Stanley Fish, Fredric Jameson, and Richard Rorty, it cannot itself be reduced to a simple summary or language.[1] Rather, as the philosopher Richard J. Bernstein argues, the postmodernist challenge to modernism is best thought of as a "constellation." By this he means a "'juxtaposed rather than integrated cluster of changing elements that resist reduction to a common denominator, essential core, or generative first principle.' . . . What is 'new' about this constellation is the growing awareness of the depth of radical instabilities. We have to learn to think and act in the 'in-between' interstices of forced reconciliations and radical dispersion."[2] This metaphor is useful because it suggests that even scholars not thought of as postmodern or who reject many of the tenets of postmodernism are still forced to deal with the issues raised within the constellation of modernity/postmodernity.[3]

It is beyond the task of this book to explore the full ramifications of the modernity/postmodernity constellation for the study of political science. Instead, we explore the implications for social regulation of two lines of argument in this literature. First, we examine the postempiricist critique of science that has a central place in this constellation. Although this critique is not exclusively the province of postmodernism, it is central to the broad questioning of the Enlightenment ideal of rationality and science as ways of knowing the world upon which much of postmodernism is based. As we contend in the next section, the Enlightenment view of science is inherent in the managerial language. Although we argue that the managerial view of science is seriously flawed, we do not reject the use of science in the policy process. Scientific knowledge, even when contested, is inevitably a part of social regulatory disputes. Our dialogic model draws upon a different view (articulated by John Dewey and others) that sees science as a particular form of discourse providing factual information useful to, but not determinative of policy-making. At the same time, our model uses the process of communication within scientific communities to draw important conclusions about democratic communities and citizenship.

The second aspect of postmodernism we find useful is its contention that there can be no final synthesis when it comes to understanding complex social phenomena. Postmodernists draw on the work of poststructuralist literary critics who argue that texts, literary or otherwise, have no inherent meaning. Instead, meaning emerges from the interaction of the reader or interpreter with the text. Every interpretation is affected by the social context within which the reader and text are enmeshed. Here, postmodernism challenges a fundamental belief of modernist thought that, despite current disagreements, if we work hard enough and long enough, a final, accurate description of the world is possible. Whatever its ultimate ontological validity, we believe that the postmodernist challenge[4] has special relevance to understanding political disputes. The contested and complex nature of the issues surrounding social regulation, especially when it affects local communities, assures that no single discourse or perspective can ever be final or completely adequate. We argue that it is ultimately futile to debate whether one discourse is better than another for capturing the so-called reality of complex political issues. Instead, we must learn to live with the ambiguity that flows from the impossibility of arriving at final agreement over the "best" or most accurate way of understanding politics.

This line of argument requires us to critique and question constantly the languages used by powerful political and economic institutions to construct and define political issues. Yet such criticism is difficult when there is consensus among powerful actors over the basic terms of political discourse. So, the limitations of the dominant managerial and pluralist discourses are easily overlooked at the national level, where there is widespread consensus among policymakers and scholars alike about the realities of political power and the best ways to understand the activities of government. Yet the adequacy of these languages is challenged by outsider groups, especially at the local level, who are not part of that consensus. Then, as many postmodernist scholars argue, only power and not reason can support policy decisions, and such decisions are exceedingly difficult to legitimate and enforce in a democratic society. What is needed is a way of thinking about policy-making and democratic politics that grants the unattainableness of synthesis and sees policy-making as a highly tentative enterprise that must incorporate multiple perspectives and languages. The dialogic model is a step in that direction.

The Managerial Perspective and the Myths of Science

As we have seen, the managerial language assumes that supposedly neutral experts can discover an optimal policy that maximizes overall utility and thereby defines the public interest. Such a view of the public interest derives from an objectivist view of truth or at least of politically relevant truth. Bernstein summarizes this view as "the basic conviction that there is or must be some permanent, ahistorical matrix or framework to which we can ultimately appeal in determining the nature of rationality, knowledge, truth, reality, goodness, or rightness. An objectivist claims that there is (or must be) such a matrix and that the primary task . . . is to discover what it is and to support his or her claims to have discovered such a matrix with the strongest possible reasons."[5] Thus, the search for correct public policies is seen as similar to the search for scientific knowledge as presented in objectivist models of science. This search assumes that there is a single answer to public policy problems, that this answer can be found within a single language, and that this language is one of scientific expertise. Additionally, the managerial language assumes that facts and values are separable and that it is the search for the former that ought to concern expert policymakers. An exploration of these assumptions helps explain the great difficulties that this discourse encounters in debates over social regulation.

At root the notion that science can provide the criteria to guide policy decisions rests on what David Collingridge and Colin Reeve call the "myth that science produces truth."[6] This view can be characterized as mythical because it is based on an idealized and inaccurate view of both science and policy-making. The managerial language assumes that science is a value-free method of getting at a kind of information that lies beyond the political conflicts that obscure and complicate policy-making. As we noted in chapter 2, this very feature of science made it attractive to Progressives in general and to the prophets of regulation in particular. However, the work of most recent philosophers and sociologists of science is at odds with the view of science that is implicit in the managerial discourse.

Collingridge and Reeve agree with Karl Popper and his followers that the nature of scientific knowledge is tentative; science does not produce final, objective truth. They argue that scientific theories are never verified as true; rather, they have simply not been falsified. If we accept the Popperian view, then the tentative nature of scientific theories and the high probability that at some point they will be modified or proven wrong make them unsuitable as exclusive guides for public policy.

Even more critical of the view of science embedded in the managerial language is what is sometimes called the postempiricist view of science. Central to this approach is Thomas Kuhn's *The Structure of Scientific Revolutions*, which, although at odds with the Popperian view of science, also explodes the myth that science produces truth.[7] Rather than progressing toward a more objective and true understanding of nature, Kuhn argues, scientific progress is discontinuous. Periods of "normal" science, which conform somewhat to the Popperian view, are punctuated by revolutionary periods during which shifts occur between noncommensurable paradigms whose differing definitions of problems and programs for research contradict the assumptions of scientific progress inherent in objectivist notions of knowledge. During such periods of paradigm change, disputes are resolved by a largely social process that is at odds with any idealized notion of objective scientific experimentation.

Following from Kuhn's emphasis on the social processes within which the scientific enterprise is embedded, many scholars argue that scientific inquiry is always shaped by social values. Some identify social values as shaping only the direction of scientific inquiry.[8] Social values, in the form of public and private funding, and the demands imposed by universities and professional associations lead scientists to investigate certain questions, while ignoring others. Labeled a contextualist approach, this position argues that the direction of scientific inquiry is shaped by social values, not by a transcendental search for truth. Contextualists argue, however, that although scientific *questions* are shaped by social values, the *answers* to these questions are not: specific scientific findings are determined only by the values of the scientific community.

Other scholars go even further, arguing that not only the questions, but the answers as well are shaped by the social context within which science takes place. Helen Longino, for example, argues that the practices of scientific communities (peer review, replication of results, and so forth) can protect against invalid results produced by individual idiosyncracies, including cheating.[9] However, science cannot protect itself from the influence of social values shared by the whole community. Thus, what passes for scientific truth is influenced by the degree to which the scientific community is open or closed to diverse views and perspectives. Longino illustrates her argument by examining changes in the way that scientific questions involving gender and race and investigated in a variety of disciplines have been shaped by the degree to which women and people of color have been admitted to or excluded from scientific communities.

Taken as a whole, postempiricist critiques suggest that science is a social process, embedded in and influenced by both the structure of the scientific community and the broader society within which that community is situated. Although science may still be the best model we have of human rationality, it is far from being the neutral, objective technique assumed by the managerial language.

Not only does science provide an inadequate base for policy-making, but the requirements of policy may produce poor science. Since most theories are ultimately proven wrong, scientific progress depends on fully investigating them. If wide-ranging investigation and creativity are to be encouraged and the problem of prematurely discarding theories avoided, the costs of being wrong must be minimized. In contrast, public policies based upon inevitably tentative scientific theories invest considerable resources in the implications of such theories and have potentially large error costs. When science is undertaken for such policy ends, it is prone to what Collingridge and Reeve call the "over-critical model."[10] Owing to the high potential error costs of theories used to guide public policy, such work is subject to premature, overly critical analysis. This, in their view, tends to stifle creativity and scientific progress. In short, scientific progress is characterized by tentative agreements among scientists over theory. These agreements are not a sufficient ground upon which to base the often life-or-death decisions confronting public policymakers.[11] This argument is especially relevant to social regulation that often couples difficult, potentially life-or-death policy decisions with high levels of scientific uncertainty and disagreement.

Following from the myth that science yields truth is the second myth: experts can be expected to agree. If experts can provide an objective source of information, then that knowledge should unambiguously lead to policies serving the public interest. Disagreement, in this view, results solely from the political biases of experts. If this bias is removed and truly neutral experts are found, or so the myth goes, then disagreement will disappear. The work of Kuhn suggests that scientific experts often disagree, especially when they are drawn from differing disciplines or from disciplines undergoing shifts in paradigms: disagreement, far from being a distortion of science, is at the heart of it.[12] Further, political bias simply cannot be avoided when experts become involved in the policy process. When operating in the policy arena, scientists act less like the neutral experts assumed by the managerial perspective and more like advocates in a court of law. Inevitably, experts view the issue through the lenses of those who have hired them.

Aside from the questionable assumption that science can offer an objective and value-free solution to policy problems, another assumption of the managerial language further limits its ability to fully capture the dynamics of social regulation. Consistent with a reliance on science, the managerial discourse is based upon a faith in progress: a belief that, on balance, the fruits of continued technological development outweigh the costs. Indeed, from its origins in the Progressive era, the discourse assumes that science can both deliver the benefits of progress and minimize the costs of continued technological development. This belief is implicit in the Progressive idea that neutral experts could develop solutions to the problems created by the negative externalities of market operations. However, because social regulation is often about the failures of science to anticipate or even comprehend such costs, it is unlikely that this same faith in progress can gain widespread acceptance as a solution to these problems.

Specific battles over social regulation turn on basic underlying assumptions about progress and science. Yet their significance is obscured by the ways in which debates are framed by policy-making institutions. As we argue below, one of the reasons that the communitarian language so effectively challenges social regulatory policies is that, in stark contrast to managerial assumptions, it is based, among other things, upon a rejection of progress and an appeal to tradition.[13]

The model of science underlying the managerial perspective has a final relevance for any analysis of social regulation. If policy-making is viewed as the domain of scientific experts whose research can produce a truth upon which public policy can be ideally based, then there is no need to consider mechanisms for allowing ordinary citizens to participate in the decisions that affect them directly. The perspective of the managerial language on political participation is well illustrated by the way it sees the role of public education.

Given the assumption that science produces correct, politically neutral policy solutions that mitigate the costs of progress, the managerial discourse views public education as a way to win support for delegating authority to experts and gaining acceptance of their decisions. Neutral experts are viewed as a superior alternative to participatory decision making, and public education is the way to convince a mass audience that this is so. Thus, increasing public faith in science becomes a means to the end of letting experts decide.[14] Yet these assumptions are difficult to sustain in areas of social regulation that necessarily involve scientific and technological uncertainty. Often, in these policies the more one knows (that is, the more educated one is), the *less* likely one is to grant decision-making authority to others, experts or not. However, the managerial discourse's notion of public education holds no solutions for the problem of how to incorporate into the policy process an educated public that still disagrees with policymakers.

The Pluralist Perspective and the Myth of Relativity

When analysts recognize the limits of the managerial discourse, they most frequently turn to a model of policy-making rooted in the pluralist discourse. Rejecting the managerial discourse on the grounds that there is no objectively verifiable public interest to guide public policy, analysts conclude that correct public policies are defined by a process that balances contending interests.

For example, in rejecting objectivist models of science, Collingridge and Reeve, like most other scholars, advocate a policy process guided by disjointed incrementalism. Such a process is rooted in and illustrates the underlying assumptions of the pluralist discourse. Owing to the inability of scientists to provide objectively truthful information, especially under the "over-critical model," any claim by experts supporting proponents of a particular policy position will be counterbalanced by the criticisms of those on opposing sides. Because the claims of experts cancel each other out, Collingridge and Reeve conclude that policy is never based upon science, and should not be.[15] Rather, given the inability of experts to produce the type of knowledge needed to guide a rational process of policy-making, coupled with the high cost of error, policy-making ought to proceed incrementally. This process allows both minimizing the risks of error (by changing policies slowly and at the margins) and maximizing the likelihood that policy will reflect a balance of opposing interests (given the usual pluralist assumptions about due process).

Rejecting the idea that science is capable of producing objective knowledge appropriate as the basis for public policy-making, the pluralist language assumes a relativist approach to political truth.[16] For pluralists, the facts of political life are always viewed through the prism of values, and substantive definitions of good public policies are entirely relative to one's political values. Further, to the extent that conflicts over values are rooted in individual self-interest, there is no way to successfully resolve value disagreements. Owing to the ultimate relativity of political values and their impact on policy positions, the pluralist view assumes that the correct public policy cannot be defined a priori and instead focuses upon process: good public policy is whatever emerges from due process. It is the role of government to establish this due process, defined as the balancing of competing self-interested groups.

Yet, *self-interest* and *due process* are terms that prove fully as problematic as managerial notions of value-neutral science and expertise. Above, we explored the inappropriateness of pluralist assumptions when applied to the circumstances of social regulation (see chapter 2). Assuring a fair compromise among contending parties is one key to successful social regulation; however, pluralist politics are unlikely to produce such compromise.[17] As with the managerial discourse, the limitations of this perspective for comprehending social regulation are rooted in its basic, and usually unexamined, vocabulary and assumptions.

One reason that the pluralist discourse cannot adequately define due process in social regulation is its implicit assumption that the policy process is, and should be, a relatively stable arena within which the leaders of established interest groups bargain to produce incrementally changing policy outcomes. At its heart, the pluralist discourse sees public policy as the vector of forces produced by competing organized interests that reflect stable, common interests among group members. These interests are articulated by elites at the heads of organized interest groups.

In effect, whereas the managerial perspective defines politically relevant truth as the neutral knowledge produced by experts, pluralism defines truth as the wisdom embedded within a stable set of public polices: its standard for politically relevant truth is the status quo. However, social regulation deals with the ignorance, or unanticipated consequences, produced by the status quo. Restricting policy options to changes at the margins may be more risky than attempting more dramatic changes if existing practices are themselves extremely dangerous (consider, for example, ozone depletion or the unsafe disposal of hazardous substances).

The pluralist discourse has even greater limitations when social regulation affects geographically defined communities. Citizens within these communities are mobilized into shifting and unstable groups that are difficult to contain within the boundaries of a pluralist policy process. These groups do not accept the legitimacy of either the stable interest groups involved in policy-making, the elites who speak for these groups, or the suitability of the status quo as a basis for incrementally changing policies. Yet more open forms of policy-making that would incorporate such groups lie outside the pluralist discourse. One reason for this inability is the way pluralists address, or rather fail to address, the process through which citizens and communities define their interests.

No less than defining due process, pluralist assumptions about self-interest are inadequate for capturing the dynamics of social regulation. Given relativistic assumptions about political truth, pluralists identify the clash of self-interest as the basic force driving public policy disagreements. Yet, how individuals and groups come to define their interests—a central question for social regulation—is exogenous to this discourse. It is simply assumed that citizens are self-interested and pursue these interests, however they define them, through their representatives in the political process. The role of government is to be simply a neutral referee that balances these interests.[18] In this respect pluralist discourse mirrors the approach adopted by free-market economists in assuming that the Invisible Hand aggregates self-interest into a social good.[19] For both pluralists and free-market economists, freedom is the absence of constraint in the pursuit of self-interest, however an individual chooses to defines it.

Such an approach may be adequate for ordering private markets, but it is problematic when applied to the political arena, especially in the area of social regulation. How individuals define wants and self-interest in the marketplace may be an entirely private matter. How they come to define their self-interest in the public arena has collective and public consequences. In *The Public and Its Problems*, John Dewey argues that the central problem of democracy is how citizens articulate their interests and then constitute themselves as a public, a problem that is increasingly important in modern societies. His argument is worth quoting at length because it uses a language particularly appropriate to the issues raised by social regulation, the conflict between community and more distant social processes, and the difficulty individuals have in understanding how these processes affect them:

> The local face-to-face community has been invaded by forces so vast, so remote in initiation, so far-reaching in scope and so complexly indirect in

operation, that they are, from the standpoint of the members of local social units, unknown. . . . An inchoate public is capable of organization only when indirect consequences are perceived, and when it is possible to project agencies which order their occurrence. At present, many consequences are felt rather than perceived; they are suffered, but they cannot be said to be known, for they are not, by those who experience them, referred to their origins. . . . Hence, the publics are amorphous and unarticulate.[20]

Dewey raises one of the key issues in forming publics in the area of social regulation: what information and calculations do citizens use to understand how broader social and economic processes affect them in their communities? Definitions of self-interest depend upon this perception. Yet by taking political self-interest as being self-justifying and outside the boundaries of politics, the pluralist discourse does not afford an adequate vocabulary for dealing with this issue.

A related problem has to do with the essential difference between public and private realms: the necessity of interest aggregation in the public realm. Individuals are free to pursue their own interests in the private market: ideally, the goods and services we purchase do not affect the goods and services others consume. In contrast, politics requires aggregating interests and ultimately the adopting of policies that affect large numbers of citizens. It is for this reason that pluralist discourse takes the group rather than the individual as the basic unit of democratic politics. Yet, at the same time, pluralism does not contain a vocabulary for specifying how common public values can be forged to guide democratic public life. Instead, in Thomas McCollough's apt characterization of pluralist theory, "Values are treated as interests when they enter the economic and political spheres."[21] Thus, by reducing even collective action to the simple aggregation of individual self-interest, pluralism delegitimizes *any* claim to, or disinterested discussion of, higher collective purposes. With respect to the impact of the new social regulation on geographically defined communities, however, the absence of such discussion is exactly what limits the ability of policymakers to justify the imposition of costs associated with regulation on specific groups and communities. This conundrum has not gone unrecognized by more sophisticated pluralist theorists. In much of his later work, for example, Robert Dahl has wrestled with the connection between public and private realms and its consequence for pluralist political theory.[22]

To the extent that values are treated as interests when they enter the public arena, pluralist discourse takes for granted the way citizens combine their self-interests into group advocacy. But the values that inform this decision have profound consequences for politics.[23] A central problem in social regulation is whether citizens see themselves as members of a wider society, with a common interest in addressing the negative externalities of technological development, or simply as members of a local community resisting the specific costs of regulation imposed upon them. These issues are obscured by taking self-interest for granted and seeing the test of good policy as the balancing of competing interests.

A fundamental problem with the pluralist discourse is its underlying assumption that the flaws in objectivist models of scientific knowledge lead inevitably to relativistic models of politically relevant truth. We disagree that policy-making need (or can) exclude scientific information, tentative as it may be, as pluralist analysts like Collingridge and Reeve suggest. When policies involve issues suffused with the complex scientific and technical questions posed by social regulation, it is impossible to exclude experts entirely from a role in the policy process. A central problem in such policies is how this information can be incorporated into the calculations used by citizens to define their self-interest and constitute themselves as engaged publics. The inability to design institutions capable of involving such experts in policy-making without having them distort the process indicates a failure of the pluralist perspective, not of science or of the possibility of incorporating technical expertise into the policy process. Responding to this failure, the communitarian perspective suggests an alternative to both objectivist and relativist models of knowledge.

The Communitarian Language and the Strengths and Limits of Localism

Taking the communitarian language seriously allows one to move past the debate between pluralist and managerial models of policy-making and develop a more adequate approach to social regulation. As with the other two discourses, the communitarian language on its own cannot fully capture the dynamics of social regulation. It is important to reject the notion that any single discourse can adequately capture the dynamics of social regulation at the local level. Yet, of the three discourses, the communitarian most fully addresses the issues raised by social regulation when it affects geographically bound communities. However, this language is either ignored by policymakers and analysts or, when addressed, is distorted and dismissed (as, for example, in the generally negative interpretation of the NIMBY phenomenon we discuss below). That is why we suggest a dialogic model that points toward a policy process that incorporates all three languages used to discuss, analyze, and debate social regulation.

The greatest advantage of the communitarian perspective is its vision of politically relevant truth, which avoids the pitfalls of the objectivist and relativist positions. The communitarian discourse and the democratic wish embedded within it assume that in a democracy political values are to be hammered out in communities. Human beings come to be human by developing an understanding of the world based upon their participation in the ongoing communication and dialogue characterizing human communities. Unfortunately, advocates of this perspective are often quite vague about their definition of *community*. Because the communitarian discourse is central to our arguments, it is necessary to provide a definition that we will use throughout the rest of the book: a community is a public entity small enough to allow, at least potentially, repeated, face-to-face interaction over issues involving both individual and collective interests. A specific political community may be defined by a residential neighborhood, a workplace, or both: we argue that the strongest democratic communities would be situated around both workplace and residence (see chapter 4).[24]

The communitarian language's alternative view of politically relevant truth is best exemplified in American intellectual and political history by the pragmatist philosophy and democratic theory of John Dewey. Dewey believed that although knowledge can never be ultimately grounded in an external reality, as objectivists claim, the existence of continuing dialogues within democratic publics establishes a framework, albeit a changing one, that avoids the pitfalls of relativism.[25] Thus, for Dewey and the communitarian discourse, democratic communities hold out the possibility of rejecting the objectivist notion that truth exists independent of human understanding, while still denying that the determination of politically relevant truth is arbitrary, as held by relativists. Unlike pluralists, Dewey rejected the view that the definition of self-interest by individuals occurs outside the political process. Instead, he argued that individuals come to understand their interests only through social interaction and association. It is this definition of self-interest, forged through public activities, that is the heart of democratic politics: communal life, for Dewey, is democracy.[26]

Bernstein draws out the parallels between this view and postempiricist views of science. As we pointed out above, Kuhn's idea of noncommensurable paradigms and the sociology of revolutionary science challenged idealized objectivist models of science. Scientific progress does not simply respond to the facts revealed by experimentation: observation is always theory-laden. However, Bernstein argues that this does not mean that it is either arbitrary or irrational, as is assumed by relativists. Instead, Kuhn's descriptions of scientific progress in periods of paradigm change, when social and psychological factors may influence individual scientists as they decide between competing and noncommensurable paradigms, constitutes a type of rationality based upon dialogue within scientific communities: "Central to this new understanding is a dialogical model of rationality that stresses the practical, *communal* character of this rationality in which there is choice, deliberation, interpretation, judicious weighing and application of 'universal criteria' and even rational disagreement about which criteria are relevant and most important."[27] Although scientific debate is over the facts, these facts do not have an existence independent of consensus within the scientific community over their interpretation.[28] When debate occurs between advocates of noncommensurable paradigms, it involves the redefinition of a consensus over what indeed constitutes the facts.[29] Truth, then, is no more nor less than what emerges from ongoing, unconstrained dialogue within human communities. Values necessarily intrude into this dialogue (thus this position does not assume the managerial myth that experts will agree), but they must be acknowledged as such, publicly defended, and denied arbitrary or privileged positions.[30] That is, in taking a position, scientists must defend themselves, and, within the parameters defined for discourse within the scientific community, there are agreed-upon criteria for judging such public defenses. The key to whether the outcomes of deliberation produce truth is the extent to which a given community establishes an unconstrained dialogue wherein the force of rational argumentation is the decisive factor. Indeed, one of the pioneers of twentieth-century nuclear physics, Niels Bohr, noted the centrality of language and consensus in modern science: "It is wrong to think that the task of physics is to find out how nature *is*. Physics concerns what we can *say* about nature."[31] This perspective navigates between the limits of both objectivist and relativist views of knowledge by

> appeal[ing] to the outcome of critical discussion by a community of in-
> quirers as the arbiter of questions about truth and reality. Truth is what
> would emerge as the result of unconstrained inquiry pursued indefinitely.
> Even our standards of evaluation cannot be established in advance, for they
> too can be improved in the course of inquiry. Inquiry, of course, never ac-
> tually comes to an end, and we can never be certain that any particular
> result will not be overturned by further investigation. But we need enter-
> tain no serious fears that our inquiries may fail to disclose things as they
> really are: the way things show themselves to our ongoing inquiries simply
> *is* the way they really are.[32]

This attempt to move beyond objectivism and relativism is more than an obscure debate among philosophers of science. It underlies the democratic wish and the communitarian discourse that challenges dominant managerial and plural-ist perspectives. In our view, more fully elaborated in the next section, the structure of scientific communities has much in common with the structure of healthy democratic communities. Community is central because it is the level of social organization best suited for structuring political institutions capable of encourag-ing individuals to consider, in addition to their own self-interest, broader questions of the public good. The interaction of individual and collective interests is a prerequisite for successful democratic policy-making. In this context, citizens can hammer out the political values and goals that will serve as a dynamic political truth.

From the communitarian perspective, communities are more than the simple aggregation of anomic individuals characteristic in liberal thought. The liberal conception is flawed because it ignores the powerful ties within communities that build collective identity. However, the communitarian view also rejects the image of the entirely organic community of much conservative thought, which entirely submerges individual autonomy in the collectivity.[33] Central to the communitarian discourse is *active participation* in political communities through which individuals come to see the relation between their self-interest, the self-interest of others, and the collective interest. These lessons emerge through repeated interaction among community members over a variety of issues involving both common agreement and conflict. Because learning to manage disagreement and controversy is an essential task of successful democratic politics, conflict, no less than consensus, is a necessary component of community building.

Such reflection on the collective consequences of decisions, best facilitated at the community level, allows mutual understanding—if not agreement. This consideration of the public good differentiates the activities of individuals in their role as consumer (that is, the economic assumption of the individual pursuit of self-interest) from their actions as citizen. However, the significance of this distinction is not well developed in either pluralist or managerial perspectives.

The communitarian discourse is especially relevant to social regulation because the effects of such policies often fall on communities, not just individuals or the entire society (the two basic categories of liberal thought).[34] When a hazardous waste facility or nuclear power station is sited, its location and operation affect not just anomic individuals or the society at large, but the particular community within which it is placed. Likewise, the imposition of pollution controls on industrial operations that threatens to close factories and alter the economic structure of a region affects the physical health and economic well-being of both residential and workplace communities. These issues illustrate forcefully to individuals the concrete connection between government action or inaction, the activities of private business, modern technology, and the well-being of one's family, neighbors, and fellow workers. It is not surprising that citizens adopt the communitarian language when challenging policies produced by institutions rooted in managerial and/or pluralist perspectives.

Key to successful social regulation is the ability of political institutions to foster public dialogue that explores the connections between individual self-interests, community well-being, public values, and the broader collective implications of public policies. In spite of the overall decline in most forms of political participation so lamented by social scientists and political commentators, when social regulatory issues arise, citizens in affected communities rapidly become mobilized and demand a role in decision making. However, existing institutions, rooted in pluralist or managerial models of state-economy relations, are unable to absorb these demands in their policy-making process. The result is the expression of citizen demands commonly referred to as NIMBY, and this leads to policy gridlock. In this view, NIMBY is far from a purely selfish local reaction; instead it results from the failure of political institutions to include newly mobilized citizens and communities into the ongoing process of policy-making.

Viewed from either the pluralist or managerial perspectives, social regulation is problematic precisely because it imposes concentrated costs on particular communities in order to benefit a diffuse public interest. Pluralist approaches have difficulty with this because such issues introduce zero–sum elements into policy-making: pluralist politics work best when compromise is possible and the benefits and costs of decisions can be allocated across many groups or individuals. Zero-sum decisions, in contrast, require policies that are necessarily unpopular with the groups forced to bear the costs, thus undermining the allocational bases of pluralist politics.[35]

Likewise, social regulation is problematic from the managerial perspective because it involves explicit trade–off decisions about values over which there is no expert or society-wide agreement (for example, jobs versus health; the value of a human life; what constitutes an adequate level of safety or an unacceptable level of risk). Such issues challenge the managerial perspective's assumption that expertise can provide the basis for rationalizing public policy. The seriousness of this challenge is evidenced by the frequent inability of experts located in bureaucratic settings either to make such policies or to implement them successfully over the organized resistance of groups that stand to bear the costs of such decisions (for example, delays in enforcing environmental regulations due to judicial appeal or failures to site hazardous waste facilities or nuclear power plants).

In contrast, from the communitarian perspective, the community-level impact of social regulation holds great promise because it opens the possibility of transcending individual self-interest through the creation of communities of common interest. Managerial and pluralist vocabularies make little distinction between self-interested individuals and the self-interested behavior of communities; both are impediments to the implementation of rational public policies. However, the self-interested activities of communities are quite different from the self-interested activity of individuals, because by definition the former require individuals to act cooperatively in the furtherance of a community's collective good. The structure of this cooperative activity is the key for a democratic politics of social regulation. Seen this way, communities are places where the powerful centrifugal forces of individual self-interest are balanced against the centripetal forces of family, church, home, and work, which bond individuals with each other in a collectivity. When public policies directly affect the well-being of communities and when this effect is apparent, there is the potential for creating within those communities the type of citizens and democratic politics needed for addressing the pressing issues of modern technological society.

While the communitarian discourse adds much to our understanding of social regulation, on its own it is inadequate for a variety of reasons. First, by focusing on a community-based definition of politically relevant truth, it fails to consider the broader national context within which social regulation takes place. Christopher Lasch argues that the communitarian vision fails to comprehend the connection between local communities and the broader political economy because it treats society and community as if they were distinct and opposed categories and thus fails to see that they are connected and inextricably intertwined.[36] If he is correct, the failure to consider the broader context is deeply rooted in a basic vocabulary that opposes the interest of the community against that of the whole. Therefore, the inability of the communitarian language to address this broader context effectively may not be easily remedied.

A second flaw in the communitarian discourse is its failure to consider the ways in which scientific and technical issues constrain the decisions of communities. At some level, estimates of the technical feasibility and the risks of regulatory alternatives need to be incorporated into political decision making. Even if these issues are seen as estimates that need to be viewed critically, it is not clear how such complex information can be incorporated into the democratic wish of direct citizen decision making. Lasch locates this failure deep in the origins of the communitarian perspective. He argues that the communitarian discourse is rooted in an appeal to past values and social structures threatened by modernity (that is, family, religion, community) that implies a fundamental critique of progress. As noted above, it is this appeal to the past that makes the communitarian discourse so attractive to those who resist social regulatory policies. However, this deep suspicion of scientific expertise and its underlying faith in progress makes the incorporation of sophisticated scientific and technical knowledge problematic for the communitarian vision of democratic decision making.

Finally, the communitarian language fails to specify adequately the institutional mechanisms for mobilizing and maintaining a citizenry capable of continuous participation in democratic decision making. As James Morone makes clear, although the communitarian language can be a potent voice for challenging existing policies and institutional arrangements, it has been much less successful at actually creating institutions and citizens capable of participating continuously in policy-making.[37] In many areas of social regulation affecting geographically bounded areas, the communitarian discourse has mobilized citizens and overwhelmed the liberal democratic institutions of local government without suggesting alternative forms of ongoing participation.

In the next section we turn to our dialogic model, which, along with the work of several democratic theorists, points the way to a synthetic model of social regulation capable of addressing the limitations of the three dominant discourses used to analyze and debate such policies.

The Role of Dialogue

Although portions of the communitarian discourse are limited, we believe that the view of politically relevant truth underlying this language provides the basis for an approach to social regulation allowing integration of all three discourses we have described. Because it ultimately assumes that truth emerges from open dialogue within communities, we call this view a dialogic model of rationality. We use this model at two levels, one empirical, the other normative. At the empirical level, we argue that in social regulatory disputes affecting geographically bounded communities, the failure to allow different discourses to confront each other, democratically, is a guarantee of continued gridlock and policy failure. As long as discourses are excluded, the gridlock so remarked upon by analysts is likely to continue. More normatively, the dialogic model and the political theorists upon whose work it draws provide a variety of insights about the characteristics of policy-making institutions capable of engaging these competing languages.

Our model focuses on four specific characteristics of social regulation affecting geographically defined communities. First, we argue that equal access to usable information is a prerequisite for meaningful participation in a democratic dialogue over social regulation. In most areas of social regulation, involving as they do complex scientific and technical issues and the distribution and redistribution of risk and uncertainty, information is a central political resource. Therefore, its distribution across contending groups must be carefully considered. Rather than employing the approach to science and expertise assumed within the managerial discourse, we base our analysis of the distribution of access to scientific information on a view of science which is rooted in the postempiricist literature we discussed above.

Second, regulatory decisions must be understood as part of a broader pattern of opportunities for citizen participation in the full range of policy decisions affecting their communities. Many of the problems of social regulation are solvable only if there is an ongoing democratic dialogue over a wide variety of issues that allows citizens to grapple repeatedly with the difficult questions posed by public life. A central problem of social regulation is that the difficult decisions such policies pose for localities are usually treated by policymakers and policy analysts as discrete issues isolated from the ongoing political life of communities. Yet there is a growing body of work, both empirical and theoretical, that supports our contention that if difficult issues of social regulation are to be solved, much more widespread opportunities for citizen participation must be designed.

Third, following from the insights of postmodernist scholars, there is a need to create policy-making institutions that can adjust to the ambiguity that inevitably flows from the inability of any single discourse to capture the full dynamics of social regulation. Bringing competing discourses into engagement, a requirement of successful policy-making, does not mean that final agreement will occur or that these competing languages can be synthesized into a single perspective. However, although final reconciliation may be impossible, these issues must be debated in public. Citizens' perspectives and the values that inspire them cannot be assumed to be outside the policy process (as is the case with the pluralist discourse), but rather must be seen as emerging from open and public debate. What is needed is a redefinition of successful policy-making that allows for the tentative and changing nature of the public interest as that concept is defined and redefined by communities affected by social regulation.

Finally, community-based dialogue over social regulation must be placed within the wider context of the political economy of federalism. Chapter 4 is devoted to this complex task.

Scientific Rationality, Political Dialogue, and the Pathology of Power

Our dialogic approach to politically relevant truth has many parallels to the views of postempiricist philosophers about how truth is produced within scientific communities. Because the communitarian discourse has difficulty incorporating the need for scientific and technical information into its approach to social regulation, it may seem ironic that its underlying view of knowledge is consistent with that of many postempiricist philosophers and historians of science. This irony stems from two different ways of viewing the relation between scientific knowledge and the policy process.

In the managerial language, scientific knowledge is a product, generated by a process external to and quite different from the actual policy process. Following its objectivist underpinnings, the managerial discourse sees this knowledge as objective and uninfluenced by political or social values. It is this view of science as product that is rejected by the communitarian discourse and accounts for its skepticism about the use of expert knowledge as a guide for policy-making. The dialogic model of rationality, in contrast, views science not as a product, but as a social process. This process, though not immune from the values and biases of the wider society within which it is situated, still provides a model for the structuring of substantively rational human communities. As such, science provides *both* information that can, under certain circumstances, be useful to policy-making *and* insights into how a democratic policy dialogue might be structured. Viewed this way, the operation of a scientific community offers instructive insights into the structuring of democratic communities.

Viewing science as a social process highlights the significance of access to the knowledge and resources necessary to challenge truth claims both for scientific rationality and for achieving substantive rationality via democratic processes. To make this comparison we first identify widespread access to knowledge and resources as a central characteristic of scientific communities that allows them to achieve rationality. Then, we ask the reader to imagine a very different sort of scientific community, one in which access to resources necessary for challenging truth claims is not widely distributed, but rather is tightly concentrated. The results of this imagination game illustrate that the distribution of access is both a form of power and a critical determinant of the substantive rationality of such communities. These insights have direct application to political conflicts involving technical and scientific issues, applications obscured by both pluralist and managerial perspectives, but highlighted by a dialogic model.

One of the features of scientific communities that allows them to achieve consensus is that results of research, even when they challenge that consensus, are stated in such a way as to be reproducible by other scientists. That is, the truth claims of scientists are, in principle, verifiable by other scientists who may question such results.[38] This possibility obviously lessens the temptation for scientists to cheat or mislead the community by distorting their research.[39] These temptations may be considerable, based as they are on the desire of scientists to produce original and counterintuitive results and theories upon which individual reputations are built. Thus, the possibility of others verifying truth claims serves to counteract and bound the pressures pulling scientists, consciously or unconsciously, toward distorting or misinterpreting their research.[40] The result is that, in general, scientists may trust the intentions, even when questioning the results, of other researchers. This trust in shared intentions is an important reason for calling the scientific enterprise a community. We return to the issue of community (both scientific and political) and its potential for mitigating the undesirable effects of the unrestrained pursuit of individual self-interest in the next section of this chapter.

Now imagine a different sort of scientific community: one in which only a single laboratory, staffed by scientists committed to a new theoretical program, is capable of performing experimental work necessary to investigate this body of theory. Scientists committed to other programs would be skeptical about the research results reported because the temptation to cheat, either consciously or unconsciously, would be significant. However, because these scientists do not have the resources to confirm such results themselves, the resulting skepticism would likely translate into a denial of any surprising empirical work and a reluctance to discard or modify accepted theory. Further, the challenge of new truth claims to existing theory coupled with the inability of those skeptical of such findings to challenge them effectively would also lead to an overall decline of trust within the scientific community.

Indeed, it might even be inaccurate to describe such an arrangement as a community at all. Such a scientific establishment would be biased heavily toward confirming the positions of those with access to the resources necessary for performing experimental work and deviate significantly from the unconstrained dialogue upon which scientific rationality depends.[41] In such a social organization, power—in the form of monopoly control over the means to conduct experimental work—and not reason would have force in determining consensus.

The same situation obtains in a substantively rational democratic community considering political issues that have scientific and technical components. The key to establishing a dialogue approximating undistorted communication is not to rely on a single set of experts, supposedly divorced from political influence; neither is it to discard the use of scientific information entirely. Rather, as with the scientific community, so too with the democratic community, all parties to a conflict must have the ability to verify or challenge independently the claims of other parties. As Walter Lippmann argued, "There can be no liberty for a community which lacks the means by which to detect lies."[42]

Such a deduction highlights the conflict between technical and substantive rationality. From the perspective of the former, the duplication of access to expertise and research entailed in guaranteeing all parties to a conflict the ability to challenge truth claims may seem a waste of scarce resources. However, from the perspective of substantive rationality, such duplication is necessary to insure that the decisions reached in a political community are the result of undistorted communication and, in addition, will be viewed as legitimate by parties to the conflict.[43]

This argument has both normative and empirical implications. At the normative level, it suggests that a central concern of democratic politics should be the equalization of access to the means to challenge scientific and technological information. But at the empirical level, it helps explain the profound skepticism toward science and experts that increasingly marks social regulatory disputes. Our argument contends that as long as policy-relevant scientific and technical information is defined by policymakers as a product generated by experts outside the policy process, citizens adversely affected by social regulatory policies are unlikely to accept as legitimate the findings of those experts unless they themselves have the means of evaluating and challenging them.

John Forester supplies a concrete example of this sort of reasoning when he notes the significance of the creation of the National Institute of Occupational Safety and Health (NIOSH), whose task was gathering information about hazardous substances in the workplace.[44] Workers were given access to NIOSH's findings, a vital source of information with which to challenge the claims of studies funded by various employers' organizations. Moreover, the legislation establishing the Occupational Safety and Health Act specified "rules structuring grievance processes accessible to workers wishing to call into question claims about the safety of work processes or work environments."[45] Putting aside the question of how these policies were implemented,[46] such arrangements are promising because they provide workers with both direct access to the policy-making process at the shop floor level and access to technical information that makes participation potentially meaningful.

A number of political theorists have drawn out the significance of this view of knowledge for the structuring of politics in a democratic society. John Dewey articulated a similar understanding of science and its relation to democracy in *The Public and Its Problems.* He argued that, in a democracy, science could not be thought of as neutral or external to the political process. Rather, the goal of science (especially the social sciences) ought be the production of knowledge and techniques directly useful to citizens. He argued that science was a form of critical reasoning appropriate to political deliberation and could and should be taught to citizens in order to guide their political deliberations. He insisted that it should not become simply a body of knowledge in the hands of elites. So, Dewey emphasized access to the means of acquiring knowledge as a prime consideration in evaluating democratic decision making.[47]

There are also strong parallels between the criteria for producing truth in this dialogic model of rationality (that is, truth emerges from unconstrained, ongoing dialogue) and the argument developed by Jürgen Habermas that democratic politics are defined by unconstrained dialogue in which only reason has force.[48] In propounding his model of undistorted communication, for example, Habermas makes an explicit attempt to incorporate a dialogic model of rationality into a theory of democratic decision making.[49] Much like Dewey, Habermas identifies communication as the basis of social activity; it is at the root of what it is to be human.[50] Thus, at the risk of oversimplification, his overall project is the development of a theory of social action rooted in language. For our purposes, the most interesting portion of Habermas's work is his analysis of what is logically implied by speech acts, or his model of undistorted communication. The ability of citizens to engage in undistorted communication over pressing public issues becomes a central requirement for democracy in Habermas's work.

The degree to which public policies and the institutions producing them further the goal of creating a self-conscious, self-reflective democratic citizenry provides an evaluative standard for policy analysts that stands as an alternative to more mainstream approaches dominated by criteria of economic efficiency and technical rationality.[51] The creation of such citizens is integral to the realization of a fully rational society. Modern societies fail to achieve substantive rationality insofar as one form of rationality—instrumental or technical—drives out deliberate consideration of higher goals.[52]

Undistorted communication is the key to achieving substantive rationality. For dialogue to be unconstrained, listener and speaker must enter into a relation aimed at achieving an understanding within which both know that the claims of the speaker are, at least in principle, verifiable by the listener. Habermas defines four requirements—or "validity claims"—for such communication, directed toward establishing understanding between individuals and communities:[53]

1. Understanding within an unconstrained dialogue implies that what is said is true. Moreover, the truth of claims made by the speaker must be, at least in principle, verifiable by the listener.

2. What is said is stated in language comprehensible to the listener.

3. The speaker is sincere or veracious with respect to his or her statements and is not trying to mislead the listener. That is, both the speaker and the listener must be committed to achieving understanding rather than manipulation. Again, this claim implies that the speaker's honesty is verifiable.

4. What is said is correct or legitimate with respect to recognized social or language conventions.

Habermas's approach emphasizes not simply the objective veracity of claims made, but rather the ability of other participants in the dialogue to verify such claims. This possibility is essential because it serves as a check on the types of claims that contending parties to a political conflict are likely to make. Access to technical information, then, becomes a form of political power. Only when such access is relatively equal among parties to a conflict can dialogue proceed along the democratic ideal wherein only reason has force.

In both scientific and political discourse, although there may be no objective way to ground any particular position independent of the participants in debate over such issues, this need not preclude *substantively* rational decision procedures. Instead, it suggests that substantive rationality resides not in the discovery of a transcendental matrix within which knowledge is grounded, but rather in the structuring of the human institutions within which dialogue over truth takes place. Here, as Kuhn argues, science is the best model we have of substantively rational human activity. If the actual operation of science is at odds with preconceived models of rationality, then it is the latter that will have to be revised, not the former. Following upon this argument, our dialogic model sees citizens as being defined politically by their role in dialogue within communities: the problem of democratic politics becomes structuring an unconstrained political dialogue among citizens.

Part of such an unconstrained political dialogue must be policy-making institutions open to a wide variety of languages and perspectives on both specific policy issues and broader procedural issues about how to make public policy. In *Beyond Adversary Democracy* Jane Mansbridge gives evidence for the importance of policy-making institutions being open to a variety of models of democracy. Mansbridge studied democratic decision making in two small communities: a New England town and an urban crisis center.[54] She uses the political processes in these communities to illuminate two models of decision making undergirding democratic theory: adversarial and unitary democracy. Adversarial democracy assumes that political issues involve fundamental conflicts of interest. In such situations, there are no objectively correct decisions (that is, there is no a priori definable public interest); rather, the central concern of democratic theory is procedural fairness. Mansbridge identifies unitary democracy as an older, but neglected model of democratic decision making. Unitary democracy assumes that the task of political institutions is pursuing policies consistent with an overarching public interest.[55]

One of Mansbridge's most valuable insights is that neither adversarial nor unitary models alone can provide the basis for successful democratic policymaking. Rather, both methods of decision making serve important functions within the same political system. This reinforces our argument that approaches to social regulation rooted in a single discourse are inadequate as the basis of a theory of democratic regulation. Indeed, a major failing of the two communities Mansbridge studied was their overreliance on a single model of decision making. In the New England town meeting, an overreliance on unitary decision making lessened the degree to which the interests of the less articulate and economically disadvantaged were considered. In the crisis center, an overreliance on adversarial methods increased the time it took to make decisions and emphasized conflict resolution without adequate attention to the role that unitary decision making could play in emphasizing common interests and consensus.

The key to structuring democratic politics is recognizing the nature of issues in terms of the degree to which they involve fundamental conflicts of interest and hence require either unitary or adversarial methods. Moreover, the identification of issues as either unitary or adversarial cannot be imposed by elites: it must involve the active participation of the citizenry. A successful democratic politics involves, Mansbridge argues, the ability to shift between these two methods of decision making.

A dialogic model poses an ideal standard that is unlikely to be completely realized in the real world. But it offers a criterion that has much the same status as the ideal of economic efficiency that underlies mainstream policy analysis. For example, many scholars argue that the market, if left unfettered by government interference, *would* produce economic efficiency, even though such a social arrangement never exists in a pure state. In comparison, we suggest that it is just as reasonable to argue that a policy process characterized by open access to information and resources for its production, undistorted by the dynamics of capitalism, would closely approximate the dialogic model of substantive rationality. This model can provide a way of evaluating political institutions as to their potential for establishing the political dialogue necessary for reaching substantively rational outcomes.

The dialogic model furnishes intriguing insights into the problems posed by social regulation. It suggests that substantively rational decision making is not precluded by the inevitable presence in social regulation of both contested scientific issues and conflicting political values. Rather, the artificial separation of scientific and political issues undermines the legitimacy of policy-making. Restoring legitimacy involves providing citizens with the opportunity to address meaningfully the political and scientific components of public policies. However, this is an enormous challenge with the most profound consequences for the structure of policy-making institutions. At a minimum, it requires the creation of a true public sphere within which educated citizens have the opportunity to understand, discuss, and influence the decisions that affect their lives. In such a public sphere, unconstrained dialogue addressing both scientific and political conflicts would decide such issues in a democratic and substantively rational manner.

In addition to providing access to information, the dialogic model requires repeated participation in policy-making issues. In spite of its other limits, a strength of the communitarian discourse is that it takes advantage of the mobilizing effect that social regulation has on communities. To the extent that citizens are drawn into public decision making because it affects the collectivity with which they have the greatest contact, such instances can provide the building block for the public dialogue necessary for democratic politics. Such opportunities, however, do not arise very often; indeed, the whole weight of mainstream expectations about policy-making, rooted as they are in managerial and pluralist perspectives, works to limit direct citizen action. The inability to incorporate mobilized groups of citizens into policy-making may be the single greatest failure of social regulation in the United States. At an empirical level, the dialogic approach indicates that social regulation will founder at the local level, unless a broader range of discourse can be incorporated into policy-making, given the availability and power of the communitarian challenge. At the normative level, rather than attempting to limit public mobilization, the dialogic model suggests that more attention must be paid to the structuring of discourse and participation in communities as citizens attempt to grapple with the daunting problems posed by social regulatory issues.[56]

The importance of taking full advantage of opportunities for establishing public dialogue may be understood by reconsidering our discussion of the parallels between scientific and political communities. If, as we argue, scientific communities provide the best model of substantive rationality, then understanding the conditions under which such communities flourish is vital. One circumstance, already discussed, is the widespread access to the means for independently verifying truth statements. Another feature of scientific communities, applicable to their political counterparts, is the repeated interaction of members over a variety of issues. Through face-to-face interaction in universities, in laboratories, at conferences, and through the written interaction of journals, scientific communities are joined with one another by a complex web of connections. Repeated contacts help to establish the shared values and mutual respect necessary for the community to survive and benefit from heated disagreement over specific issues.

Repeated interaction in a variety of settings is no less important for political than for scientific communities. Taking advantage of every opportunity for involving individuals in public community dialogue is necessary to establish the shared values and mutual respect needed to sustain communities through disagreement over specific political issues. Interesting applications of game theory by Robert Axelrod suggest that, in a variety of political contexts, repeated interactions between opposing parties increase the value placed on future outcomes. This leads to shared norms and values that, in turn, produce cooperation, compromise, and reduction in the intensity of conflict.[57] These findings are quite similar to Mansbridge's argument that engaging in unitary decisions helps communities build the shared norms and respect necessary to successfully negotiate more divisive adversarial procedures. Over time, shifting between methods of decision making would generate the repeated interactions, in a variety of settings, necessary to create the shared norms and values required to allow communities to weather disagreement over specific issues. Only within such ongoing communities would it be possible to maintain the continuous public dialogue central to vital democratic community life.

Such a sustained political dialogue (and the shared norms and values it produces) holds the promise of a truly democratic politics capable of achieving substantively rational political decisions. However, for such norms of cooperation and conflict to develop, citizens must be regularly involved in public dialogue about a wide range of issues. If participation is limited to a small number of highly charged issues, the "evolution of cooperation" described by Axelrod is unlikely to occur. It is the general paucity of opportunities for citizens to participate in political decisions that makes the rare instances when they are mobilized so conflictual.

It is, however, naive to believe that, by themselves, citizen groups mobilizing around specific local issues will produce the ongoing, undistorted political dialogue required for substantively rational decisions. Significant institutional reform is first required in order to provide a public sphere capable of fostering democratic dialogue within which community groups can operate. One of the major limits of the communitarian discourse and its democratic wish is a failure to specify the mechanisms for creating a continuously engaged citizenry. Until reform occurs, the absence of participation or informed citizenship, in general, cannot be taken as a sign of either consensus or apathy. Here, we agree with Benjamin Barber that citizens are apathetic because they are powerless, not powerless because they are apathetic.[58] In fact, Barber's *Strong Democracy* gives valuable insights into how policy-making institutions might be structured to encourage the continuing participation of citizens in decisions that affect their communities.

Strong Democracy

We find the approach to democratic theory developed in *Strong Democracy* especially relevant to our dialogic model for two reasons. First, Barber's democratic vision is based upon the creation of a citizenry capable of participating in a public dialogue that would address the issues posed by social regulation. Central to our dialogic approach to social regulation and to Barber's strong democracy is the insistence that the creation of enlightened citizens is essential, but it cannot be achieved around a single issue (especially one as divisive and difficult as, for example, hazardous waste regulation), but rather depends on the creation of broad-based opportunities for participating in public decisions. Policymakers who attempt to develop support for their regulatory policies by educating and mobilizing citizens over very specific issues are likely to fail unless they can situate their effort within the context of much broader attempts to establish public dialogue. The existence of this sort of ongoing dialogue is especially crucial when the contending parties operate within the logic of alternative discourses. It is only in the context of the give and take of ongoing public dialogue that the competing assumptions made by such groups can be clarified and the grounds of disagreement identified and narrowed. Significantly, Barber is useful because he is quite specific about the institutional changes required to create a citizenry capable of participating in a public dialogue that would address the issues posed by social regulation (we return to the specifics of these institutional reforms in chapter 8).

Second, Barber defines the public interest in a way that is especially compatible with our argument that no single discourse is capable of capturing the dynamics of social regulation, and therefore policy-making institutions must be capable of living with the ambiguity and tentativeness that flow from this postmodern dilemma. His idea of strong democracy opens a path to the creation of a democratic politics of social regulation that would allow citizens to determine for themselves, through ongoing public dialogue, the adequacy of competing conceptions of public policy and the public interest. The conditions for strong democracy parallel the problems posed by social regulation: "Strong democracy is politics in the participatory mode where conflict is resolved in the absence of an independent ground through a participatory process of ongoing, proximate self-legislation and the creation of a political community capable of transforming dependent, private individuals into free citizens and partial and private interests into public goods."[59] Barber stresses the need for forums at which citizens come together to talk publicly about the issues that affect their communities: this rooting of citizenship in community dialogue—or "democratic talk"—is the basis for democratic governance.

Especially appealing is his argument that direct citizen participation must lie at the heart of a true democratic politics; here he implicitly assumes a dialogic model of politically relevant truth. Moreover, his focus on the institutional preconditions for meaningful, as opposed to symbolic, participation and dialogue parallel the logic of Habermas's model of undistorted communication.

Barber argues that the simple allowance of participation is not sufficient; the conditions for meaningful citizenship must first be created. Thus, he does not take participation for granted, as do those operating within the pluralist perspective. Indeed, he explicitly rejects the "thin" vision of democracy underlying pluralist models. Rather, he recognizes that private market systems of production coupled with liberal democracy create structural impediments to meaningful participation. Unless specific measures are taken, the relations of domination, subordination, and inequality found in the private sector are likely to be reproduced in the public sector. In particular, Barber is critical of the tendency of elites to allow public participation (usually in the form of oversimplified public opinion polling) in only the most difficult of public issues.[60]

Three conditions must be met for democratic talk, or public discourse, to develop: first, citizens must have the means to acquire the information they need to make decisions; second, they must have the power to make decisions; third, public discourse must be ongoing. The first requirement parallels the central characteristic of Habermas's undistorted communication model and of our discussion of scientific and political rationality: equal access to resources necessary to verify independently truth claims made by contending parties.

The second requirement is often overlooked: establishing dialogue is not sufficient; needed too is the power to implement the outcomes of dialogue. Public dialogue, no matter how structured, remains purely symbolic unless it is connected to the institutional means for action. Attempts to establish an informed polity through the use of new technologies (for example, interactive cable television stations, local access stations, CSPAN) fail to engage more than a tiny segment of the population because they do not connect the dissemination of information or participation in debate to any control over policy outcomes. From this perspective, offering education without empowerment does not provide incentives sufficient for motivating individual citizens to give up competing private pursuits and to invest the resources required to enter into public debate over vital issues: Barber writes, "Give people some significant power and they will quickly appreciate the need for knowledge, but foist knowledge upon them without giving them responsibility and they will display only indifference."[61]

The third requirement connects the idea of the public interest to the dialogic model of truth. Public discourse must be ongoing, not simply a one-shot occurrence, in order to allow decisions to be tentative and subject to revision. This is crucial because, for Barber, the public interest is not something to be discovered, but rather created in an ongoing process of discourse that builds citizenship in addition to making public policy: "Political willing is thus never a one-time or a sometime thing . . . , but an ongoing shaping and reshaping of our common world that is as endless and exhausting as our making and remaking of our personal lives."[62]

This treatment of the public interest highlights the political implications of a dialogic model of truth. Within this framework, the public interest is not something that can be objectively discovered, as is assumed by the managerial approach, nor is it an entirely relative concept without substantive meaning, as is assumed by the pluralist approach. Rather, in moving beyond objectivism and relativism, the dialogic model asserts that truth comes from ongoing dialogue; it is not an end product that can be defined once and for all. Thus, the political truth embodied in the public interest is something developed, changed, and improved over time through a continuous public dialogue that constitutes democracy. In our view, the central question for democracy in a modern society is whether it is possible to establish communities that allow for citizens to participate meaningfully in such public dialogue.

Can the Dialogic Model Work: Is It Possible/Necessary to Establish Open Public Discourse?

An objection to our argument often raised by scholars operating within either the pluralist or managerial perspectives is that only a small proportion of the population can ever be expected to possess the desire or sophistication to participate in public discourse. Indeed, they argue that a well-functioning pluralist political system requires that only a small proportion of the polity be mobilized at any given time; therefore, we need not worry about the lack of political sophistication of the masses and their general failure to take part in politics.

Such arguments, whatever merits they might once have had, are rendered moot by recent changes in the dynamics of American politics: changes that are very much in evidence in social regulation. Increasingly, elected policymakers tailor their words and actions to the results of public opinion polling, reported almost daily in the mass media. Further, by publicizing and dramatizing certain political issues, the media also help to arouse the concerns of many citizens.[63] Here, the modern mass media and public opinion polling perform a distinctly populist role as they constrain the actions of elites by reporting and shaping the opinions of the masses. Whether this populist role is or can be consistent with the requirements of democratic politics is an important question. This is the dilemma raised by the increasing influence of new media like talk radio. Nevertheless, although only a small proportion of the population might be politically mobilized or sophisticated, the ability to tap mass public opinion instantaneously and report it makes the ability of average citizens to discuss and think through the political issues of the day critical to contemporary American politics.[64] The frustration of political elites with these changes was illustrated recently when President Clinton assailed most Americans for being "political couch potatoes." Given these dramatic changes in the mass media and in American politics, perspectives that assume that the impact of the masses on policy issues can somehow be limited or that increasing the political engagement and sophistication of the polity are not crucial issues seem much more utopian than the work of theorists like Barber, Dewey, or Habermas.

The unreality of elite pluralist assumptions is especially apparent when one examines social regulatory issues as they affect local communities. In such cases the severity of the problem coupled with media attention has mobilized large numbers of formerly uninvolved citizens. In the absence of an ongoing public dialogue capable of fostering compromise and enlightened participation, these groups use existing institutions (for example, administrative appeals procedures and the courts) to simply stop the policy process in its tracks.[65]

Conclusion

We believe that the dialogic model has much to offer the study of social regulation. By transcending the limits of pluralist, managerial, and communitarian languages, this perspective provides the basis for a fresh analysis of the political issues lying at the heart of social regulation. At the very least, the model of dialogic rationality provides a critique of regulatory policy-making and the limits of most reforms aimed at remedying failures of policy-making institutions. To the extent that the policy process as well as proposals aimed at its reform is based upon untenable assumptions about the nature of science, its role in the policy process, and the relation between political power and access to the resources necessary to challenge truth claims, substantively rational social regulation is unlikely. At an empirical level, then, the dialogic model provides a set of predictions about the dynamics of social regulatory issues affecting geographically defined communities. By emphasizing these issues, policy analysis based upon a dialogic model (and the democratic theory that flows from it) is a much needed alternative to analysis rooted in either the pluralist or managerial perspectives.

In addition, democratic theorists who employ a dialogic model of rationality provide more than a critique of existing institutions. They also develop a blueprint for the ways in which substantive rationality and the undistorted communications upon which it is based might be actually achieved through the creation of democratic institutions organized around democratic communities. This is especially important in social regulation when contending interests operate within the confines of unexamined but competing discourses. In such cases, public dialogue that can narrow the range of disagreement and clarify contested concepts and assumptions seems an essential prerequisite for successful policy-making.

However, theorists who operate within this perspective remain vulnerable to a central limitation of the communitarian discourse by focusing too exclusively on the individual or the local community. Although we too see the community as the primary location for dialogue among citizens, we also argue that any practical reforms aimed at establishing undistorted communication must take into account the larger political structures within which communities operate. In particular, the nature of social regulation makes it essential that the structural dynamics of federalism, which make a strong role for the national government a requirement of any successful reform program, be analyzed. That is, because it would require a substantial redistribution of political and economic power for strong democracy and dialogue at the community level to be realized, the federal government must act as the guarantor of such political processes. In the next chapter we complete our theoretical treatment of social regulation with a structural analysis of the impediments to participation and dialogue posed by the dynamics of a federal political system operating within the confines of a capitalist economy.

4 / Democracy, Redistribution, and Social Regulation in a Federal System

Understanding social regulation requires not only an examination of the languages used in public debate, but also an appreciation of the national structures of capitalism and federalism within which such policies are made. Given our focus on language, attention to structural context is important for two reasons. First, as most theorists whom we draw upon make clear, discourses cannot be analyzed solely at a rhetorical level because they are systematically connected to patterns of political and economic power. Yet, a danger of focusing on political language is that rhetorical analysis sometimes stops short of analyzing the relation between discourse and structures of political and economic power. In this chapter we explore this relation.

Second, given our concern with the impact of social regulation on geographically defined communities, the relations between local actions and the broader national context within which social regulation is formulated and implemented must be understood. This connection is often overlooked by reformers who tend to focus on either national or local solutions to the problems of social regulation. Daniel Press, for example, argues that a problem with implementing many of the political reforms advocated by environmentalists is that such reforms tend to advocate either centralist solutions (in which great power is granted to the central government for specifying societywide regulations enforced everywhere) or decentralist solutions (in which almost complete autonomy is granted to localized communities adopting environmental solutions contingent on the specific conditions facing that community).[1] We contend that the reforming of social regulation requires an approach that goes beyond the grand either/or of national-centralist/local-decentralist solutions. Such an approach involves political institutions that are capable of engaging the disparate languages used in debates over social regulation and a structural analysis that can specify the results of creating such institutions at various levels of government.

Our structural analysis focuses on two crucial features of social regulatory policy in the United States. First, most social regulations have the effect of redistributing income. Second, regulations are promulgated and implemented within a federal system that distributes responsibilities among various levels of government. In this chapter we argue that the interaction of these two features places significant limitations on the policy-making capabilities of state and local governments.

In addition to presenting a theoretical explanation for the unfolding of social regulation, our arguments here have important consequences for the creation of institutions and citizens capable of fostering the public dialogue central to the dialogic model of policy-making. In chapter 3 we concluded that meaningful public dialogue requires that citizens be empowered and drawn into the policy-making process to a much greater extent than is currently the case. However, thus far we have focused primarily upon reforming community-level political institutions as key to democratic policy-making. This emphasis on the local level and a tendency to ignore the broader context of both federalism and capitalism within which such reforms would have to take place reflect the limits of the communitarian discourse and the assumptions of many democratic theorists, especially those whose work bears upon the dialogic model. A broader perspective is necessary because assigning greater policy-making responsibility to the local level, in an attempt to foster democratic governance, is politically complicated. In a market economy, where control over investment is in private hands and both political and economic power is unequally distributed, the allocation of authority among levels of government for policy decisions with redistributive effects is never politically neutral. Rather, the very assignment of authority to one level of government as opposed to another has dramatic implications for the interests of contending political and economic groups as well as for policy outcomes. Likewise, shifting responsibility always changes the balance of power among these interests. This chapter poses a larger context for our study of social regulation by addressing these issues.

The distribution of control over capital, across both groups and regions, has wide consequences for the development of social regulation. In particular, the private control of capital severely limits the sorts of redistributive policies that can be undertaken by local and state governments. If citizens are to participate in meaningful public dialogue about social regulation, this dialogue cannot be created by local governments or communities acting on their own.[2] Maintaining public dialogue that is open to the perspectives of all three languages of regulation requires a persistent, yet transformed role for the national government, precisely because social regulation raises fundamental questions about economic and political power at the state and local levels. Essentially, social regulation changes the rules of the game in local and state policy-making (see chapter 2), and we argue that the powerful presence of the national government is necessary to ensure that the "new rules" are recognized in the policy process.

Our task in this chapter is twofold. First, we examine the structural characteristics of a federal system to help us explain and predict the development of social
regulation in the United States. We test these predictions more formally as part of
our analysis of hazardous waste regulation in chapter 5. This analysis provides the
structural context for our case studies of the uses of the competing languages of
social regulation presented in part II. Second, we explore the normative implications of our theory for reforming regulatory policy-making consistent with the
prescriptions of the dialogic model. In particular, this chapter's focus on the
politicoeconomic context within which regulatory policy operates allows us to
specify the redistribution of authority among levels of government that would be
required to foster public dialogue and a true democratic politics of social regulation. This focus also allows us to evaluate the political and economic impact of
other current proposals for reforming social regulation. Combined with the arguments of the previous two chapters, our structural analysis constructs a framework
for evaluating the pitfalls of current social regulatory policies and for explaining
why these policies are so often seen as illegitimate when attempts are made to
implement them at the local level.

Regulation, Redistribution, and Federalism: A Structural Analysis

Earlier we noted the redistributive effects of regulatory policies and the significance of this for the politics of regulation (see chapter 2). Because they alter
market-determined distributions of income, all regulatory policies necessarily take
income from one group and redistribute it to other groups. Social regulation
usually attempts to redistribute income from producers and consumers to large
latent groups adversely affected by the negative externalities associated with production and consumption activities. Because producers and consumers will mobilize to protect their incomes, regulatory goals are often difficult to achieve.[3]
Reflecting the silence on this issue of the three languages used to analyze and
debate social regulation, the literature on regulation largely ignores the consequences of multiple levels of government—except as general barriers to coordina-
tion and implementation—and no one has analyzed explicitly the relation between
federalism and the redistributive effects of such policies.[4]

Scholarship on intergovernmental relations and federalism, however, demonstrates that a federal system poses serious structural problems for the implementation of redistributive policies. This literature is rooted in what is usually called the
public choice perspective—an attempt to apply the insights and assumptions of
microeconomics to the study of politics. From this perspective, the multiple,
autonomous governments in a federal system have the decided advantage of allowing citizens to choose among the alternative mixes of public goods and services

offered by state and local governments. Government ceases to be a singular and systemwide monopoly supplier, and, as a result, citizens can choose among the various public policy offerings of competing jurisdictions. The possibility of citizens "voting with their feet" forces governments to be efficient and responsive to the demands of citizens because they can move from one jurisdiction to another. In short, from the public choice perspective, the advantage of federalism is that it can potentially create "quasi-marketplaces" among competing jurisdictions, thus providing many of the advantages assumed to flow from the operation of private markets.[5] Moving between empirical and normative analysis, public choice theorists have been vigorous defenders of multiple small, overlapping jurisdictions and opponents of attempts to create larger units of government pursuing more uniform public policy goals.[6]

Although more recent scholarship questions the degree to which citizens actually do vote with their feet (and thus force local jurisdictions to offer responsive bundles of public goods and services),[7] one widely accepted conclusion of the public choice approach is that local levels of government have great difficulty implementing redistributive policies. This is so because those citizens who stand to lose from such policies always have the option of exiting to another jurisdiction. When the wealthy and/or politically powerful are the groups whose income will be reduced by such policies, it becomes structurally impossible for local politicians to impose losses. Public choice theorists tend to conclude from this that redistributive policies are the province of the federal government because at the national level it is quite difficult for those who are adversely affected to escape negative policy consequences.

A serious limitation of this perspective is that it remains firmly rooted in a pluralist discourse. Uncritically accepting the supposed advantages of a private market economy, public choice theorists assume that individuals are all politically free to choose between the various mixes of goods and services that are offered by competing governments. They tend to ignore both the constraints on political and economic freedom imposed by the unequal distribution of political power and economic resources as well as the origins of such distributions. In particular, the ability to exercise the "exit" option or the impact of the threat of exit is not equally distributed throughout the population. Public choice theorists, however, seldom consider the relation between redistribution and federalism.

An exception to this general trend is Paul Peterson's *City Limits*, which is a more complete exploration of the public choice approach. While Peterson accepts many of the assumptions of the public choice approach, he also explicitly acknowledges and seeks to incorporate the work of Marxist scholars who focus on the role of capitalism in the functioning of a liberal democracy. He analyzes systematically the degree to which control over capital conditions the responsiveness of government to various segments of the polity at different levels in a federal system. His arguments serve as the departure point for our own approach to the structural implications of federalism for social regulation that affects geographically defined communities.[8]

Peterson argues that most research on politics in a federal system fails to grasp different characteristics and capabilities of different levels of government: city and state governments cannot be analyzed as if they were national governments. Instead, state and local governments must be distinguished from national governments because they are relatively limited in the types of policies they can pursue. These limits derive not only from the more obvious constitutional and legislative limits on autonomy, but also from the intense interstate and interurban competition for economic development.[9] This competition is ultimately driven by the private control of decisions about investment and the inability of states and localities to regulate the flow of either capital or labor across their borders. Thus the combination of capitalism and federalism places great power in the hands of those who control capital relative to state and local government officials.[10]

Peterson orients his theory around three different public policy arenas distinguished by their differing impact on the local economy. Developmental policies "strengthen the local economy, enhance the local tax base, and generate additional resources that can be used for the community's welfare."[11] Redistributive policies rely on the revenues generated by more wealthy taxpayers to provide goods and services aimed at those who pay little in taxes and hence "help the needy and unfortunate and . . . provide reasonably equal access to public services."[12] Allocational policies fall between developmental and redistributive policies and consist of "marginal expenditures for services [that] have neither much of a positive nor much of a negative effect on the local economy."[13]

States and localities are structurally constrained to pursue developmental policies and avoid redistributive policies. Developmental policies are designed to attract those individuals or corporations that will expand the economy and create a growing tax base, thereby providing greater resources for public goods and services without increasing tax rates. Because states and localities cannot compel those private individuals or corporations who control investment decisions either to remain in or to migrate to particular jurisdictions, public officials must offer an attractive mix of public services to induce private investment and migration decisions favorable to their jurisdiction.

Redistributive policies, in contrast, must be avoided by states and localities because they provide services for those who pay proportionately less in taxes by taking from those who pay proportionately more.[14] If adopted, such policies would give the latter an incentive to leave the community and relocate in an area where they receive higher service levels for their taxes. A central consideration for states and localities is that relatively low exit costs make such movements a real possibility. Thus, the inability to regulate authoritatively the flow of labor and capital across borders locks states and localities in an interjurisdictional competition to attract industry and middle-class households that might bolster their tax bases. The exodus of high-rate taxpayers that might be triggered by the vigorous pursuit of redistributive policies would work against the overriding goal of economic development.

Peterson's analysis helps connect our discussion of political language to the structural dynamics that underlie policy-making. His work defines the limits of the pluralist discourse for describing policy-making at the state and local level. Peterson argues that pluralist politics—in which policy outcomes are established by the tugging and hauling of organized groups competing in the public arena—occurs only in allocational policy-making. There, contending groups struggle over the relative size of their portion of a fixed public budget. In contrast, the managerial discourse dominates developmental policies, which are made by elites (public officials and business leaders) outside the realm of open public discussion. State and local political leaders consider their attracting of new investment and middle-class residents to be noncontroversial politically because these initiatives expand the entire budget to accommodate more allocational policy-making, thereby satisfying all interests in the pluralist fray. Because they are assumed to be in the interest of all residents and therefore above politics, elites use the managerial language to discuss developmental policies. Yet, especially at the local level, the use of this language obscures the inequitable aspects of public decisions by ruling out open debate over the degree to which developmental policies actually benefit all citizens.

Similarly, Peterson helps us understand the limits of the communitarian language as it is used by groups otherwise excluded from the local policy-making process. In spite of persistent demands from disadvantaged groups, state and local officials are limited in their capacity to pursue redistributive policies. They respond to these groups in two ways: (1) symbolically, by passing legislation but not funding its implementation and enforcement; and (2) by funding such policies out of existing revenues, such that redistributive demands are added to the allocational agenda. Significantly, neither response entails actual redistribution of income because government officials fear the prospect of the exit of capital to a less burdensome competing jurisdiction.

Not all jurisdictions are equally constrained from pursuing redistributive policies. States and localities that have the soundest economies (that is, those that pursue the most successful developmental policies) are freer than poorer jurisdictions to respond to demands for redistribution. Politicians in wealthy states and localities are able to use the surplus generated by a bountiful economy to meet the needs of economically less powerful groups without dramatically increasing either the costs borne or the benefits received by corporations and middle-class taxpayers. Ironically, only those states and localities that are successful in the long run at diverting the demands of the disadvantaged can accumulate the fiscal capacity to address in limited ways a redistributive policy agenda.

By highlighting the problem of the implementation of redistributive policies, this argument has important implications for federal policy-making. When the federal government relies on states or localities to implement policies with redistributive effects, these policies will be resisted by officials at those levels. Local policymakers have every incentive to avoid using federal funds for policies that have significant redistributive impacts. To the extent that these funds can be shifted to developmental purposes (even if this means thwarting the express purpose of federal policies), individual states and localities can improve their competitive position relative to jurisdictions that might employ these funds for redistributive ends. Therefore, redistributive policies place a special onus on the federal government for strict enforcement and oversight lest they be altered by state and local governments.

This argument is structural and does not depend upon the ideology of state and local political officials. That is, no matter what the ideology of political decision makers, interjurisdictional competition, driven by the private control of investment decisions, leads them to pursue economic development and avoid redistribution. Politicians may defy this logic by undertaking redistributive policies; however, they will be punished by economic decline (caused by their inability to attract or retain industry and middle-class residents) and, ultimately, by electoral defeat, which results from their being held accountable for economic decline.[15]

Peterson's perspective is important because he, unlike most public choice theorists, identifies the degree to which the supposed freedom of choice and governmental responsiveness provided by a federal system is not shared by all segments of the population. This unequal distribution of political power belies the assumptions of the pluralist perspective. Rather, state and local governments are most responsive to those segments of the population that control private investment decisions. The mobility of capital across state and local political boundaries drives the interjurisdictional competition and, in turn, defines the structural limits of state and local policy-making. These structural limits cleave along class lines and mean that public policy will not be responsive to the demands of many groups, no matter how large or well organized at the state and local levels, if these groups call for policies that are inconsistent with the preferences of those who control capital investment.

Although Peterson contributes to our understanding of the relation between capitalism and federalism, he fails to develop fully the consequences of his arguments. He clearly identifies the structural biases that prevent local government officials from responding to demands for redistribution, and he identifies the class biases inherent in such structural dynamics. However, like most public choice theorists, he concludes that because redistribution is impossible at the state and local levels, it is more appropriately taken up at the federal level. But Peterson does not consider how the biases he identifies at the local level affect the federal government's policy-making. The relation between capitalism and federalism not only makes redistribution unlikely in states and localities, but also works to thwart the development of such policies at the national level.[16] Here, Peterson's analysis encounters the limits of the pluralist language in which it is couched.

This two-sided relation is central to our analysis of social regulation because it illustrates the limits of the communitarian discourse as a guide to comprehending and structuring social regulation in geographically defined communities. Grass-roots groups that advocate redistributive policies, like many forms of social regulation, tend to be organized, if at all, only at the local level. Indeed, those groups representing the interests of poor citizens are the least likely to possess the resources to be organized, and the likelihood of their being organized diminishes at the state and national levels of government. Thus, there is an irony in the relation between capitalism and federalism: the interests of grassroots groups are articulated at precisely that level of government that is incapable of responding to their demands. Politicians who attempt to champion the causes of such groups at the state or local level are faced with the structural disincentive of choosing policies that will disadvantage (at least in the short run) the economic health of their districts.

The Structure of Local Politics: Residence versus Workplace

In Peterson's view, the limited political voice of disadvantaged citizens at the local level ultimately depends on the necessary distinction drawn by local officials between political and economic issues at the local level. However, the inability of local groups to challenge this segregation is also shaped by the history of class relations at the local level in the United States. Although Peterson only hints at this, the connections between class divisions, political power, and the limits of local governments in a federal system are more fully developed by Ira Katznelson in *City Trenches*.[17] His work is especially consequential because it analyzes explicitly the historic connections between federalism and capitalism that must be considered in any attempt to establish democratic dialogue at the local level.

Katznelson analyzes the significance of the traditional separation of citizen and worker in local politics. Mirroring the more general Madisonian fear of the threat that mass democracy posed to private property, he argues that at the local level, the early extension of suffrage was made less threatening to the interests of capital by focusing workers' political energies toward their neighborhoods and the distributive policies characteristic of patronage party organizations. In this way, newly enfranchised workers could participate in politics (thus legitimating the state) without redistributing the power of capital through the ballot. Political participation developed as a complex but fragmented competition among neighborhood-based ethnic, racial, and religious groups over a limited range of benefits to be distributed by political machines. Thus, by defining the nature of political participation narrowly, the politics of residence fragmented the potential of class voting in a decidedly pluralist way.

Peterson agrees with Katznelson's conclusions:

> The political machine was an institution marvelously suited to the needs of businessmen in a rapidly industrializing society. . . . Strikes, unionizing efforts, socialist agitation, violent confrontations between workers and police, and systematic suppression of union leaders by corporate detectives and federal judges were regular features of late-nineteenth-century politics. But these factory disputes seldom had a decisive influence in local politics. Although local political parties sometimes cooperated with union leaders, they were never beholden to them. Political machines seldom put their weight behind the most vigorous expressions of working-class protest; by and large, they stood on the side of "law and order" in the local community.[18]

Thus, as the labor movement matured, Katznelson notes, "at work, workers were class conscious, but with a difference, for their awareness narrowed down to labor concerns and to unions that established few ties to political parties."[19] The advent of trade unionism cut across ethnic, religious, and even racial lines, but organized workers around the narrow issues of wage and job security, in exchange for the tacit recognition by management and the state of the unions' legitimate presence.[20] In effect, then, the obvious conflict between labor and capital at work was defined not in terms of class power, but in organizational terms of compensation and security, to be negotiated by elites in a managerial context that preserved the vested interests of organized labor and capital in the status quo of industrial expansion.[21]

From the dialogic perspective, the separation of residence and workplace in local politics is accomplished by the encasing of the politics of residence within the language of pluralism, emphasizing the bitter struggle of neighborhoods over the distribution of shares of the status quo: thus Katznelson's characterization of urban politics as "trench warfare." Meanwhile, labor politics, having conceded control of the workplace to management, becomes nothing more than labor economics, a managerial discourse over wages among national elites in the twin hierarchies of labor and management.[22] The differing pluralist and managerial perspectives underlying citizen and labor politics, respectively, simply do not allow for easy discourse between the two. This separation makes local politics-as-usual peculiarly incapable of grasping the problems of power and inequality characteristic of the communitarian approach.

Peterson calls this "low pressure politics," in which the limited ability of local government to address issues that are meaningful to the general public renders the formal means of participation irrelevant to political activity, and "formal channels of participation and communication are the mechanism upon which workers, minorities, the unemployed, and the poor are particularly dependent. . . . In their absence, access to decision makers is more likely to be reserved to the economically prosperous, the socially prominent, and the bureaucratically influential."[23] Peterson makes this argument resignedly, claiming that the undemocratic nature of local government keeps it from being distracted by "things political" (that is, redistributive issues) that it cannot resolve, given its resource constraints, and encourages it to engage primarily the issue of economic development. However, controversies over social regulation that affect communities concern the ways in which many forms of productive activities that generate jobs and increase the tax base may pose hazards to the well-being of residential neighborhoods. In doing so, these issues challenge the inability of local governments to address such issues as well as the historical segregation between the politics of workplace and residence.

The Structure of Federalism: Horizontal versus Vertical Mobility

Another result of the relation between capitalism and federalism also tends to be overlooked by Peterson. Like most public choice theorists, he considers the mobility of capital strictly in terms of the ease of its movement between cities or states competing for economic development *within* levels of government in a federal system. We call this horizontal mobility. But the superior organizational resources of capital allow it to move its influence more easily than underrepresented groups *between* levels of government as well, in order to avoid redistributive public policies. We call this vertical mobility.

In the rare case in which a visible crisis (such as an environmental accident) and / or entrepreneurial politics inspires government policy with truly redistributive consequences, economic interests may exploit their vertical mobility and bring pressure to bear on the system of federalism itself to thwart the implementation of redistributive policies. When corporate interests face redistributive policies at the state or local level of government, they often take advantage of the relative weakness of local groups at the national level of government to advocate a relocation of policy-making powers to the national government, in order to preempt state and local initiatives. Conversely, when the source of redistributive policy is the national government, corporate interests can (1) attack the implementation of redistributive policy in the federal agencies, where their lobbying strength is highly developed; and (2) support and exploit regulatory delegation of federal authority to state and local government as a means of allowing the redistributive policy to wither away in a governmental context structurally adverse to its implementation.

There is, then, a certain irony to federalism that has great relevance to the study of social regulation and the dialogic model we developed in the previous chapter. Citizen groups are most likely to organize at the local level, where environmental issues are visible and touch their communities and it is easiest to overcome the barriers to collective action (recall that we identified community as a key pathway between individual self-interest and sense of public purpose). However, the political power of these groups is most distorted by federalism at precisely this local level.[24]

First, when such groups resist redistributive policies (as in the case of grassroots groups that object to the siting in their communities of hazardous waste facilities or other undesirable projects), their power will be magnified—to the extent that they represent middle-class communities—by the inability of local officials to impose redistributive costs on so-called desirable taxpayers. The result is either the failure to build such facilities anywhere (as with the current problems of permitting nuclear power plants) or a pattern of siting facilities near poor black communities, a practice that is often termed environmental racism.[25]

Second, when citizen groups organize to impose redistributive costs on industry (as in the case of right-to-know regulations), their power is limited by the inability of local officials to counter the persistent threat of capital flight. Even if they succeed in forcing through policies with redistributive effects, two undesirable consequences often result. First, to the extent that redistributive policies disadvantage the jurisdiction in its attempts to attract investment, the locality will be punished for such policies by a declining economy and tax base. Second, should such policies become widespread and acquire the support of many states and localities, powerful groups that stand to lose from such wide enforcement may exercise vertical mobility to preempt them by shifting the locus of policy-making to the federal level. While the national level is the level theoretically most equipped to pass and enforce redistributive policies, it is also the level at which groups demanding redistribution are most poorly organized.

Implementation of the Resource Conservation and Recovery Act (RCRA) during the 1980s is an example of what we call vertical mobility. In spite of general support for reducing federal regulatory control and increasing state autonomy, the Reagan administration reversed its position when the state of Washington pressed for more stringent enforcement standards in the cleanup of the Department of

Energy's Hanford Nuclear facility. In the words of Richard Harris and Sidney Milkis,

> One cannot help noting the impressive irony of these administrative actions in a presidency as fully dedicated to reducing the role of the central government. The heavy-handed intervention of the Department of Justice in the RCRA negotiations over the Hanford site . . . seems to abandon a strong commitment to federalism. At a minimum it shows that the concerns of a state received relatively short shrift when they conflict with conservative principles. It is difficult to see how, in the abstract, this situation differs from a Democratic administration's intervening in state affairs to impose some liberal principle such as affirmative action.[26]

The arguments articulated to this point have implications for the unfolding of social regulation in a federal system. They postulate a structural context for analyzing the ways in which the competing languages of social regulation operate in the case studies of hazardous waste regulation presented in chapters 5 and 6. Fundamentally, the assignment of responsibility for promulgating and implementing social regulations in a federal system must always be considered a political issue in and of itself, an issue that profoundly affects the balance of power between contending groups that bear the costs and benefits of regulatory policies. When social regulation has a significant redistributive component, as it usually does, enforcement will be resisted not only by the specific interests subject to regulation but also by state and local officials. Thus, when states are given the responsibility of enforcing regulations with redistributive effects, they will seek to evade implementation. State resistance should increase in proportion to the economic significance to the state of the regulated groups. However, states with healthy, diversified economies should be more likely to implement such regulations because such jurisdictions would be insulated from threats to economic development that might be leveled by regulated groups. Further, if economically powerful, well-organized groups were to lose in policy struggles at the state and local levels, we would expect them to seek preemption of local authority by attempting to shift the locus of regulatory policy to the national government.

An important consequence of this argument is that the federal government inevitably plays a large role in the success of social regulatory policies. For such policies to succeed, federal oversight must be close and persistent. When federal attention lags, state and local officials will tend to underenforce policies by responding to them ineffectively or symbolically. Although federal support is a necessary component of successful regulatory policy, this does not mean that the current structure of the national role in such policy-making is adequate. In the next section we use the dialogic model to explore alternatives to current patterns of federal participation in the process of social regulation.

Rethinking the Federal Government's Role in Social Regulation

Our analysis suggests that the redistributive impact of social regulation imposes severe structural limits on state and local governments and that these limits must be considered seriously in any effort to reform the policy-making process. This is especially so when social regulation affects geographically defined communities because in such policy disputes citizens focus their demands on a level of government that is severely limited in its ability to respond. So, while the participation and public dialogue called for by the dialogic model are most easily achieved at the local level, their effectiveness is likely to be minimal unless the limits of local government are addressed.

The dialogic model, when considered in the context of our discussion of federalism, heightens awareness of the specific limits of community politics in American politics. Responding to communitarian calls for participation, citizens find that the place in which they can most easily organize is the place in which, owing to the structural arrangements of federalism, much of what they want cannot be accomplished. A large part of the frustration and cynicism that currently characterize citizens' attitudes toward government and government officials flows from this feature of policy-making. Given the structural limitations of state and local governments, effective public dialogue can occur only if the federal government acts to overcome the obstacles to organization, information gathering, and effective participation in policy-making that face citizens affected by social regulation. It is toward this task that federal efforts need to be directed.

Failure to address these issues leads to the situation that exists currently whereby the limited nature of local politics leads to policy debates that cannot address the redistributive issues underlying social regulation. On the surface these constraints create the appearance that citizens, through apathy or lack of opposition, accept the managerial- or pluralist-inspired solutions to such policy problems. Yet such appearance of consensus dissolves and the inadequacies of the policy process are revealed when implementation is foiled by the vigorous resistance of the communities upon whom costs fall. Attempts to implement such decisions provoke adversely affected citizens to organize and participate in nontraditional ways (that is, through NIMBY groups) that reinforce the structural inability of policymakers to impose redistributive policies on any but the weakest groups. The failure of existing policy-making institutions to incorporate NIMBY groups effectively into the matrix of state and local politics leads to gridlock and exposes the inability of these institutions to deal with many of the basic issues raised by social regulation.

At minimum, the reform of social regulation must consist of more than a simple assigning of greater responsibility to low levels of government. This alone is an important conclusion because there is an ironic convergence of the arguments of neoconservatives and grassroots reformers for giving greater regulatory authority to states and localities. Albeit in very different ways, both argue that decisions made at these levels are more democratic and efficient because they are made by governments that are close to and responsive to citizens and other interests affected by regulatory policy. What such calls fail to address explicitly is that, given the redistributive impact of social regulation, such a shift is not politically neutral. Rather, the structural dynamics of a federal system mean that such reforms would increase the power of business and other well-organized interests to resist the redistributive costs of regulatory policies, while limiting the ability of disadvantaged citizens to further their interests in regulatory disputes.

This does not mean that we reject calls for greater local responsibility. Indeed, our whole argument rests on the need to bring competing solutions into engagement. Part of this engagement means rejecting the either/or approaches of centralization or decentralization of policy-making responsibility. Our dialogic model suggests that the encouraging of political participation at the local level, where the connection between individual and collective interest is most apparent, is a building block of citizenship. Citizens' critical faculties are most likely to be developed at this level and over these issues. Ways must be found for open and fair dialogue to take place at the local level.

Neither do we reject the neoconservatives' argument that "command and control" regulation, involving the setting of specific requirements by federal agencies, is often inefficient. Indeed, such standard setting is especially problematic when it occurs as an unfunded mandate and enforcement funds and oversight are not also provided. However, if such decisions are to be made fairly at the state and local levels, the distribution of political power must be addressed. Unless the federal government works to redress imbalances in political power, the delegation of social regulatory authority to states and localities called for by conservatives is little more than an attempt to advance the interests of business masquerading as a democratic rhetoric of placing government closer to the people.

Somewhat paradoxically, then, increasing local responsibility for social regulatory decisions requires a continuing strong role for the national government. However, the nature of this role must be substantially altered. In general, to increase local autonomy in regulatory policies affecting geographically defined communities, the federal role must be changed from one of setting national standards to one of guaranteeing empowered participation in decision making at the local level. The federal role needs to be redefined as one not of establishing "correct" regulatory standards and outcomes, but of guaranteeing fair access that includes providing resources that would allow all citizens to participate in meaningful, substantive public dialogue. Therefore, just as it rejects managerial logic, so too this notion of fair access must transcend pluralist languages that fail to address the structural impediments to participation and information gathering at the local level. Thus, the federal role must include aggressive efforts to balance the systematic differentials in access to information, autonomous expertise, and political power between groups engaged in regulatory disputes at the state and local levels.

As we have seen, the national government's role in social regulation is usually conceived within the framework of the managerial discourse. Both critics and defenders of a strong federal role in social regulation assume that the predominant role for the federal government is to set regulatory standards that apply to all states and localities. This line of thinking reflects the managerial belief that there is a correct regulatory policy best determined by federal agencies. As we saw in chapter 2, this approach to the federal role in social regulation also finds expression within the pluralist discourse as interest group liberalism, that is, its belief that fair policy emerges from interest group struggles surrounding the rule-making process in federal agencies.

Reforming social regulation along the lines envisioned by our dialogic model requires challenging the dominance of the managerial and pluralist discourses in thinking about the national government's role. First, reflecting the inadequacies of the managerial discourse, social regulation, despite its technical and scientific components, is at root a political issue for which there is no a priori definable correct policy. Second, reflecting the limits of the pluralist language, if policy is to reflect the entire range of interests, it cannot simply be made by groups organized and influential at the federal level. Rather, as the communitarian discourse and the democratic wish emphasize, policy decisions must involve the direct participation of citizens affected by regulatory decisions.

Consistent with the dialogic model, we define social regulation affecting geographically defined communities as an essentially political issue in which the public interest can be determined only tentatively by public dialogue. Such dialogue requires a policy-making process that is open to all three discourses used to debate policies. Especially important is the incorporation of the communitarian discourse, which is typically ignored but is necessary to address fully the issues raised by social regulation at the community level. However, effective incorporation of the communitarian discourse used by grassroots groups requires more than the passive and sporadic opportunities for public hearings currently employed in social regulatory policies. Indeed, the failure of such hearings to overcome the structural inequities we have outlined in this chapter often leads to an increase in citizens' dissatisfaction with the entire policy-making process.

We argue that government efforts must actively and broadly work to overcome obstacles to effective, ongoing participation by citizens in the social regulatory issues that affect their communities. This would necessitate a fostering of ongoing citizen participation at the local level even if decisions might ultimately be finalized at other levels of government. Such involvement would give citizens experience in dealing with technologically complex *political* issues.

Following from our discussion in this and the last chapter, we conclude that federal social regulatory policy needs to foster participation in ways that would allow citizens to make for themselves many of the regulatory decisions that affect their communities. These efforts need to revolve around three specific concerns quite ignored in current national endeavors to set standards. First, following from the dialogic model, national policy must concentrate on creating *ongoing* participation in social regulation and not simply sporadic, piecemeal participation in only the most heated and controversial regulatory issues. Second, efforts must focus on remedying the dramatic inequity in the generation of and access to the scientific and technical information that is central to regulatory decision making. Finally, reflecting the concerns of this chapter, efforts must be made to equalize the power at the state and local levels among groups struggling over social regulations that impose or avoid redistributive impacts on geographically defined communities.

Having defined this new role of the federal government in social regulation, we find that precedents for such federal activities do exist. We have already discussed one such example, the creation of NIOSH, which is assigned to gather information about hazardous substances in the workplace.[27] This agency gave workers a valuable source of information with which to challenge the claims of studies funded by various employers' organizations.[28] NIOSH, then, helped equalize the generation and distribution of information by affording workers direct access to health and safety data at the shop floor level, where mobilization was most likely to occur. Access to NIOSH research insured that workers' participation in policy-making could be more meaningful and effective as well.

We can discover further instructive precedents for federal policies designed to achieve the goals we outline if we consider areas other than regulatory policy. In many other arenas the federal government attempts to overcome the structural impediments to organization and information gathering that face structurally disadvantaged interests at the local and state levels. Such models must be considered if the delegation of policy authority to state and local governments is to serve the democratic goal of allowing citizens a greater role in the social regulatory decisions that affect their communities. We conclude the chapter by exploring three such federal policies not usually thought of as being relevant to studies of social regulation: enforcement of voting rights legislation; the Community Action Program (CAP); and the Legal Services Corporation (LSC). Envisioning a new role for the federal government would start with a serious exploration of such policies and their relevance to the creation of policy-making institutions that are capable of effectively involving citizens at the community level in the ongoing process of defining the public interest with respect to social regulation.

Federal attempts to enforce the Voting Rights Act of 1965 are relevant to reform of social regulation because they represented attempts by the national government to redistribute political power in states and localities. Thus, they provide a precedent for federal concern with defining what constitutes open and fair access by citizens to policy-making at lower levels of government. As with social regulation, federal intervention was necessary to ensure that political decisions (that is, elections) in states and localities would not mirror the unjust distribution of political power at those levels of government. Further, in the initial enforcement of the Voting Rights Act, the focus of federal efforts was on reforming the political process rather than mandating specific political outcomes.[29]

Most relevant to our concerns, federal action was a response to the recognition that the structure of state and local governments in the South made them unable or unwilling to afford African-Americans open access to the ballot box. Yet federal efforts did not focus, at least initially, on mandating specific election outcomes (that is, requiring the election of black officials). Rather, intervention aimed to create open access to the ballot and thus a fair electoral process. Therefore, efforts were aimed at ensuring African Americans' participation on a fair and equal footing. Once the conditions for fair participation were achieved, the specific outcome of elections was left to citizens themselves.

Of course, as Abigail Thernstrom demonstrates in *Whose Votes Count?* fair and equal access is a difficult term to define.[30] Beyond simply opening the polls, federal officials and the courts, as states and localities devised election rules and drew jurisdictional boundaries that diluted the impact of newly enfranchised black voters, soon had to decide what constituted an equal vote. Such determinations necessarily changed over time as African Americans became a more and more powerful force in many states and the contours of southern politics changed. Consistent with the dialogic notion of the public interest as tentative and changeable, Thernstrom shows how shifting political realities made any hard-and-fast determination of federal intervention impossible.

Similarly, in social regulation, federal efforts would be necessary to realize the goal of the dialogic model: insuring that structurally disadvantaged groups (that is, those who call for regulatory policies with redistributive impacts) are able *both* to enter into public debate and to have their discourse, especially the communitarian language, influence the outcome of regulatory decisions. Although determination of what constitutes effective voice would be fully as problematic as it is in the area of voting rights, it is clear that an open debate over such matters and the redistribution of political power that would flow from it cannot occur without the involvement of the federal government.

What must be avoided is restricting the federal role to mandating specific regulatory outcomes that must be achieved by all states and localities. Here again, the history of voting rights enforcement is applicable. As Thernstrom argues, federal efforts ran aground when they shifted from attempts to guarantee open access to the electoral process to a focus on achieving specific outcomes (that is, insuring that a certain number of African American or minority officeholders were elected). The latter was an easier goal to operationalize and enforce, but it shifted attention away from the more crucial question of how to define effective political participation by minority citizens. Many of the same dilemmas are posed in federal attempts to develop social regulatory policies. Although it is easier to mandate specific regulatory outcomes, the more fundamental issues raised by social regulation involve questions of effective citizen participation, questions that can be addressed only by overcoming structural impediments to ongoing participation in decision making at the state and local levels.

Should the attention of federal policymakers actually turn to such concerns, there are policy institutions that might be studied to suggest the shape that future efforts might take. We discuss two such examples: the CAP and the LSC. Both programs have received their share of negative attention, scholarly and otherwise, as part of the War on Poverty, and we argue that much of that negativity has been inspired by the programs' efforts to redistribute political power at the local level, efforts that transcended both managerial and pluralist understandings of the problem of poverty.[31]

One of the issues that concerned the architects of the Johnson administration's War on Poverty programs was the necessity of overcoming the lack of sustained political participation by poor citizens in America's cities. Summarized in Daniel Patrick Moynihan's much quoted, much misunderstood, and much maligned phrase "maximum feasible participation," one goal of the War on Poverty was to increase the direct participation of local citizens in shaping federal antipoverty programs in their communities. The theory behind such efforts was that the best way to improve the plight of poor citizens was to increase their political organization and power at the local level, thus enabling them to enter the political process on their own.

One specific means of creating this empowerment of poor citizens was federally funded but locally run CAPs, which were designed to allow residents in poor neighborhoods to plan and implement the federal programs aimed at assisting them. Indicating the substantial power of the federal government to alter local power structures, these organizing efforts proved "too successful" in many cities as CAPs organized protests and demonstrations against the actions of local government officials. Exploiting their "vertical mobilization," these officials were able to lobby their congressional representatives to limit the autonomy of local CAPs. Increased funding for the Vietnam War and the reluctance of President Johnson to raise taxes to pay for both guns and butter ultimately made the War on Poverty in general and CAPs in particular victims of federal fiscal exigencies.

Although CAPs were an extremely controversial component of the War on Poverty, they exemplify the types of institutions and resources the federal government might bring to bear to aid citizens who are disadvantaged by the structural biases of local politics and allow them to gain effective access to the policy-making process. Indeed, the very controversy that swirled around these programs helps one appreciate the strong opposition at the state and local levels faced by groups advocating policies with redistributive impacts. This opposition also reveals the obstacles to building policy-making institutions that are open to the communitarian discourse of grassroots groups: such endeavors are unlikely to succeed without aid from higher levels of government.

In spite of the ultimate inability of CAPs to sustain federal support, their history reveals that even brief federal efforts can yield impressive results in altering the structure of local politics. Many scholars suggest that the CAPs conferred valuable experience on a whole generation of inner-city political leaders in poor minority neighborhoods and left in their wake the rich network of neighborhood organizations that currently exists in many American cities.[32] Yet these organizations, even when successful, have become simply another set of interest groups competing for a share of fixed local resources. As we argued above, it is precisely this transformation of redistributive demands into allocational policies, subsumed within the discourse of pluralist politics, that defines the limits of local politics in addressing the redistributive impact of social regulation. True public dialogue at the state and local levels is possible only when the federal government is prepared to maintain an ongoing presence to overcome the impediments to organization and participation faced by local groups in communities forced to bear the adverse consequences of social regulatory policies. Given the realities of pluralist politics-as-usual at the state and local levels, the continuing effective participation by citizens central to the dialogic model is unlikely without a strong, determined federal presence.

The failure to develop and support ongoing participation results in more than just the inability of citizens to have a say in the social regulatory issues that affect their communities. Sustained participation is a prerequisite for the development of norms of cooperation that allow communities to address the zero-sum issues involved in many policy disputes. Absent these norms of cooperation, citizen participation is sporadic and reactive, which leads to protests that prove capable of stopping the implementation of unpopular policies but incapable of solving the problems posed by social regulation. For that to occur, citizens must be able to enter regularly and meaningfully into the policy process, and that will happen only with the support of the federal government. Thus, there is an essential, unresolvable tension that must be recognized between the desire for local decision making and the necessity of consistent and strong federal involvement.

When social regulatory policies adversely affect local communities, this suggests that one goal of federal intervention in local decision making might be to insure the continued involvement of citizens—especially those from poor neighborhoods, which are most likely to be the target for the location of undesirable land uses—in siting and managing any facilities built in their communities. Opportunities for continued involvement would need to be supported through aggressive organizing and recruitment, perhaps based on the CAP model. Indeed, attempts to maintain ongoing participation in the management of facilities, supported by federal requirements for citizen recruitment and inclusion, might help diffuse the intense conflict that typically occurs over such unpopular facilities.[33]

Two other areas require federal attention if an open dialogue on social regulation affecting local communities is to be established: equalization of access to information and reduction in the ability of state and local interests to exploit the low exit costs of movement across political jurisdictions in order to forestall adequate discussion of the redistributive impact of social regulation. Here again, should federal efforts in social regulation turn in this direction, models are available. Of particular interest is the Legal Services Corporation, another War on Poverty program created in 1966 to provide indigent legal assistance. The focus of LSC was on injecting federal resources into local communities, in this case to balance the inequalities in access to the legal system facing poor citizens:

> [LSC] allocated federal funds for indigent legal services. Storefront law offices opened in the communities' poverty-ridden neighborhoods. Community residents selected some members of the programs' governing board of directors. Perhaps most important, the [LSC] emphasized the goal of employing the legal system to reform laws and practices perceived as biased against the poor. . . . Further, legal services programs assisted poor communities in developing their own political and economic resources by helping the poor to organize as groups and advising newly created and extant poverty organizations.[34]

Unlike CAPs and many other War on Poverty programs, LSC has survived, and in fact increased in size dramatically throughout the 1970s. Funding increased from $90 to $321 million between 1975 and 1981, and by 1981 LSC included 323 local agencies, 6,200 lawyers, and 3,000 paralegals and served more than 1 million clients per year.[35] This does not mean that the program was uncontroversial. Indeed, it attracted critics from both the left and the right. The left accused LSC of being less effective than its rhetoric of "law reform" suggested. The right saw it as a radical challenge to the legal status quo. Conservative antipathy to LSC turned it into one of the Reagan administration's favorite targets, and funding declined throughout the 1980s. Nevertheless, the program is interesting because it indicates that sustained federal efforts designed to redistribute local political power are possible.

Moreover, LSC is an example of how the federal government might equalize access to technical and scientific information about social regulation, a central requirement of the dialogic model. A component of LSC was the creation of a nationwide network of backup centers that supported community legal clinics by serving as clearinghouses for legal research in a variety of specialty areas (for example, worker's compensation, poverty law, tenant-landlord relations). In this way, local clinics were able to share information and learn from one another.

We find this an intriguing model for the ways in which technical information about specific social regulatory issues might be made available to citizens' groups, thus giving them resources for challenging the expert advice available to opposing groups (recall again that the ability to challenge claims by competing interests is central to the dialogic model). Similar federally funded but citizen controlled clearinghouses could prove a valuable resource to citizens as they become involved in social regulatory issues in their communities. Such federal intervention might help to incorporate into the communitarian language the scientific and technical information so necessary to a balanced understanding of social regulation.[36]

Federal support for providing information across localities might also lessen the distorting effect on state and local regulatory policy of the threat of exit. Offering local groups not only technical information, but also information about regulatory struggles in other communities might make exit less possible for industries seeking to escape the regulatory policies of states and localities. That is, knowing about businesses' reactions to regulatory policies in other communities would help prepare grassroots groups for attempts by these companies to move into their own communities. If such businesses knew that they would face preexisting opposition in communities to which they might move, they would be less likely to exercise the threat of exit.[37] In the absence of federal policies that equalize access to information, it is unlikely that citizens or state and local governments on their own will be able to establish the open dialogue necessary to legitimate the social regulatory process.

Conclusion

Here we conclude our theoretical discussion of the issues raised by social regulation when it affects geographically defined communities. Because none of the three languages discussed here can articulate fully the technical, political, and economic issues raised by social regulation, gridlock and citizen hostility frustrate policymaking and implementation. In our view, overcoming gridlock and frustration requires the nurturing of open public dialogue capable of incorporating the vocabularies of all three languages of social regulation. Yet this is an extremely complex task, involving as it does the creation of citizens capable of participating in a democratic dialogue about issues that directly affect their communities.

In spite of the limitations of communitarian language and the fact that it tends to be ignored by scholars and policymakers alike, we find intriguing possibilities for the realization of our goals in the way grassroots environmental groups used this discourse. However, any consideration of social regulation as it affects local communities must take into account the structural context of federalism within which state and local governments operate. This means that, whether based upon the communitarian language or not, an open public dialogue at the local level cannot be established without the active support of the national government.

In the next three chapters, we ground our theoretical discussion by applying it to the development of a specific area of social regulation affecting geographically defined communities: hazardous waste policy. Our case studies allow us to explore empirically an area of public policy-making that illustrates the challenge to the state and local governments posed by social regulation. In particular, they test the ability of state and local governments to confront the redistributive implications of social regulatory policies and their failure to address adequately the concerns of citizens, especially in the siting of disposal facilities.

We show how our theoretical approach helps to further understanding of the ways in which this critical area of social regulation has developed, why there is such widespread dissatisfaction with hazardous waste regulation, and the steps that need to be taken if this dissatisfaction is to be overcome. In short, we argue that struggles over hazardous waste regulation and other social regulatory policies need to be understood not simply as specific and technical issues, but rather as struggles about much broader questions raised by the attempt to construct a democratic citizenship in a modern, technologically advanced society.

PART II / Hazardous Waste Policy: Regulatory Failure and the Grassroots Response

We chose hazardous waste regulation for empirical analysis to illustrate (1) the limitations of the pluralist, managerial, and communitarian languages of social regulation, (2) the importance of direct citizen participation in regulatory policy-making, and (3) the presence of structural constraints on the implementation of redistributive regulation in a federal system. The brief history of hazardous waste regulation in America illustrates each of our points in ways that make its failures easier to grasp and the prospects for reform more susceptible to evaluation.

In chapter 5, we discuss the development of federal hazardous waste regulation, the states' capacities to respond in general to the regulatory agenda of the federal government, and finally, the specific efforts of three states to regulate hazarous waste. In chapter 6, we chronicle these states' struggles with siting hazardous waste disposal facilities (HWDFs). Chapter 7 connects the failures of state hazardous waste regulation with the mobilization of citizens who challenge these policies by asserting their right to know about the chemical hazards in their communities.

Hazardous waste regulation is classically within the realm of what we have referred to as the new social regulation insofar as it attempts to redress the consequences of improper toxic substance disposal—a "market failure" in which hazardous waste generators avoid paying the full price of safe disposal in the short run by shifting those costs and the attendant long-term risks to the general public. The intent of regulation, then, is redistributive, aimed at forcing waste generators to absorb the (concentrated) cost of safe disposal in order to benefit (diffusely) the general public. The redistributive consequences of hazardous waste regulation required a strong dose of entrepreneurial politics to overcome congressional reluctance to legislate in this area. But entrepreneurial politics also has its disadvantages: although it can inspire passage of legislation, the inspiration often wanes in the enduring efforts necessary to implement and enforce its mandates. The result is often a distortion of the original intent of the legislation, or, worse, government efforts become mere symbols with no real impact upon the legislation's target. Hazardous waste regulation turns out to have been no exception to this rule.

But the redistributive consequences of hazardous waste regulation make it exceptional in other ways. It asks not only those generating hazardous waste to bear the costs of its management; but it also redistributes the risks, costs, and benefits disparately among the general population. The impact of hazardous waste policies is local as well as national. When federal and state investigators identify previously unknown hazardous waste sites for cleanup, residents nearby become aware of increased health risks and often endure the very real costs of declining property values. When government approves the location of HWDFs, the decision concentrates risks and costs borne by a small, identifiable group of people and institutions (typically residents near the site and local governments providing services for the site) and diffuses benefits among the general public.

Hazardous waste regulation, then, is a form of social regulation that dramatically affects geographically defined communities. The intensely local nature of this redistribution finds local political institutions entangled in the web of American federalism. Their political capacity, historically shaped by the pluralist discourse surrounding allocational policies, is no match for the redistributive conflict that HWDF siting engenders. The language of pluralism cannot reconcile the general benefits of waste management with the local resistance to its consequences in the targeted community.

The pluralist language is not the only one available. Policymakers also define hazardous waste in the managerial language of science and technology. This allows them to discount the importance of public participation in the discussion of the inevitably political consequences of hazardous waste regulation. An elite of so-called experts typically emerges to dominate the regulatory arena. As a result, the locally redistributive impact of regulatory policy is also defined scientifically and technically—with two consequences. First, the barrier of expertise excludes the general public, particularly at the local level, from entering into official debate or having a meaningful impact on the consequences they will bear, thus violating the participatory aspects of both the pluralist and communitarian discourses. Second, because the data and assumptions of the scientific and technical arguments are often incomplete or questionable, the debate itself is characterized by uncertainty, which undermines the credibility of experts so essential to the managerial language of regulation.

The resulting policies and their redistributive impact are seen by affected publics as arbitrary because the presumption of objectivity cannot be maintained in an environment of uncertainty. Experts confronted with uncertain factual information inevitably fall back upon value judgments to fill in the gaps. The myths of science cannot disguise the fact that these judgments are innately political; yet value judgments remain unacknowledged within the managerial frame of reference.

Inevitably, hazardous waste regulation becomes political because it is redistributive and because it involves political judgments as well as factual determinations. The combination of these characteristics cannot be legitimated within either the pluralist or the managerial language; nonetheless, these languages have established the existing boundaries for understanding and structuring the regulatory process. The resulting mismatch between politics and process produces the crisis in social regulation that we noted more generally in chapter 2. Exacerbating this crisis is the fact that hazardous waste regulation must be implemented within a federal system. States attempt to transform its redistributive burdens into allocational ones either by funding hazardous waste regulation from general revenues (thus diffusing the fiscal burden to the general population) or by neglecting to regulate effectively or at all (thus allowing the social costs—and industry profits—of unsafe toxic disposal to accumulate).

The crisis in hazardous waste regulation manifests itself in several ways, and one of the most obvious is the unwillingness of citizens to bear their part of the redistributive burden involved in implementing hazardous waste policies. Because their substantive input is ignored in shaping hazardous waste policy and because they are asked to bear burdens for which their experience in political participation does not prepare them, citizens resist decisions that affect them in predominantly negative ways. They mobilize to protest the siting of HWDFs, thus producing what their critics label the NIMBY response to HWDF siting. We argue that this response is not simply the self-serving reaction of residents unwilling to pay the price for the benefits they receive from modern technology. Rather, it is a symbol of the public's deep distrust of the regulatory apparatus charged with solving the hazardous waste problem. In chapter 8, we employ the normative insights of the dialogic model to suggest the shape of policy-making institutions that are capable of overcoming this distrust through the creation of a more democratic politics of social regulation.

5 / The Politics of Hazardous Waste Regulation

Federal Hazardous Waste Policy: A Chronicle of Regulatory Distrust

In this chapter, we develop the political context of hazardous waste regulation, focusing on the events that contributed to government's incapacity to regulate hazardous waste effectively and to the subsequent demise of public trust in government policies. Our inquiry begins with the federal government's first efforts to address the problem of hazardous waste in America and follows those efforts through their implementation within the federal bureaucracy during the 1980s. Following this is a quantitative examination of states' readiness to execute those programs at the time they were charged with the task. We conclude with an in-depth examination of three very different states' decade-long attempts to regulate hazardous waste.

Following the evacuation of families from their homes near Love Canal, New York, in August 1978, Congress seriously examined the hazardous waste disposal problem for the first time.[1] Investigators estimated that there were 50,000 hazardous waste disposal sites across the United States. More than 750,000 hazardous waste generators were producing an estimated 57 million metric tons of the waste annually.[2] To make matters worse, the EPA concluded that only 10 percent of that waste was being disposed of safely.[3] These figures alarmed the public and government officials in light of the dramatic and frightening occurrences at Love Canal. More telling for our concerns is the comparison of these figures with an EPA estimate of 1987 that had annual hazardous waste generation in the United States exceeding 275 million metric tons, nearly a fivefold increase in less than a decade.[4] The Government Accounting Office (GAO) issued a report in early 1988 estimating the existence of as many as 425,380 potential hazardous waste sites—with a "high likelihood" that 130,340 of those were contaminated with hazardous waste.[5]

No one argues that the generation of hazardous waste increased that much in such a short time or that so many new sites had been created. But this comparison illustrates a central theme of our analysis in two ways. First, government knew very little about the magnitude of the hazardous waste problem as it embarked upon regulating it. Second, the definition of what constitutes hazardous waste has had an arbitrary effect upon the perceived size of the problem—the more inclusive the definition, the "larger" the problem.

Both of these points represent part of the prevailing uncertainty surrounding hazardous waste regulation, an uncertainty that makes information about hazardous waste subject to dispute and to exploitation by adversaries in the policy process. Uncertain information discredits managerial justifications of regulatory policy; and the adversarial nature of pluralist approaches to policy-making encourages the exploitation of uncertain information by opponents in the policy process. In a controversial setting, the uses of such information are likely to lead to public disenchantment with the capacity of government to regulate the problem effectively.

Public reaction to the hazardous waste issue—and the entrepreneurial politics associated with that reaction—represent a second theme of our analysis. The public received the news of the hazardous waste problem in a curious fashion. In spite of the salience of other environmental issues in public opinion during the decade prior to the discovery of the Love Canal site, hazardous waste was not a concern. Public opinion polls did not even measure the level of public awareness about hazardous waste. Passage of landmark federal legislation regulating hazardous waste and toxic substances in general occurred in 1976 with little fanfare.[6] But by 1980, the problem topped all other pollution issues in public opinion polls.

Heightened public awareness of the problem was based upon little detailed knowledge about hazardous waste and came with a reluctance to live with the results of "safe disposal." In 1980, only 22 percent of those surveyed knew what actually happened at Love Canal, but 50 percent objected to living within even one hundred miles of a safe and properly monitored HWDF.[7] In its formative stages, public reaction to hazardous waste was marked by alarm at the extent of the problem, an incomplete understanding of the issues involved, and a fundamental distrust in government's ability to respond to that problem.

From the perspective of our dialogical model of policy-making, the presence of all these factors suggests the need for at least three initiatives: first, public education about the problems of hazardous waste as a preliminary to, rather than justification for, policy-making; second, a joint commitment on the part of government and industry to document thoroughly the extent of the problems with adequate resources provided to challenge the claims made by interested parties; third, substantial involvement of citizens in the generation of appropriate solutions. Not surprisingly, these efforts never materialized.

A New Social Problem

In large part, government's lack of data on the problem and the public's lack of interest prior to 1978—and its generally uninformed suspicion thereafter—were a result of the nature of the hazardous waste problem itself. Industrial and consumer pollution has always been hazardous[8] at some level, but wastes have become increasingly hazardous with the introduction and widespread use of petrochemicals after World War II.[9] The character of certain classes of petrochemical wastes has particular relevance when one considers its unique nature in the waste stream. First, the threat that these wastes pose to human health and the environment is often discernible only in the long run, and then sometimes only tenuously. Second, the harmful long-term effects of these substances is facilitated by their extended toxicity. Third, their toxicity and long-term effects can be established only through expensive, painstaking research that involves both laboratory and epidemiological investigation. Finally, the research supporting the rapid development and introduction of new chemical compounds for commercial distribution has generally been proprietary and essentially unregulated by government until recently. Research thus tended to be secretive and often neglected to explore harmful secondary effects of new products as they were rushed to market.[10] All of these features beg the question of safe disposal practices. But industry and government simply followed standards of disposal suitable for less hazardous wastes without knowledge of or regard for the new threats posed by these chemicals or their wastes.

The incentives to dispose of hazardous wastes unsafely were enormous;[11] and government, uninformed and unmotivated to consider the long-term social costs of conventional disposal practices, never attempted to correct the market failure that these practices represented. Industry, consumers, and government essentially mortgaged the social cost of safe hazardous waste disposal as a temporal externality, not factored into the price of the products at the time of manufacture and sale.

By the 1970s, that mortgage came due. Environmental awareness, technical advances in monitoring and detecting chemicals in the environment, research relating chemical exposures to human disease, and new forms of government monitoring began to reveal the consequences of past disposal practices. More and more dumpsites were discovered to be leaking—contaminating ground and surface water and the air, and, of course, threatening human health. Repeated disclosures of newly discovered hazardous dumpsites across the country were followed by revelations of illegal, or midnight, dumping and the involvement of organized crime in hazardous waste disposal. The market for modern industrial products had created a new environmental crisis.

The Government's Response

The government's response at this point was crucial—first, in terms of mitigating the dangers of decades of unsafe disposal practices and, second, in terms of informing and reassuring the public that the dangers of hazardous waste would not go unchecked. On the one hand, addressing hazardous waste problems would have redistributive consequences, and this fact worked against effective government intervention. On the other hand, the time was ripe for the emergence of entrepreneurial politics to exploit public support for a legislative solution to the hazardous waste problem. Government, at both the national and state levels, responded haltingly to the regulatory challenge.

Several related factors contributed to the government's dilatory response. First, much of the legislation aimed at addressing the problem was conceived before either Congress or the general public was aware of the magnitude of the hazardous waste problem. Second, the implementation of that legislation, always the weak point in entrepreneurial policy-making, occurred at a time when the notion of regulation itself was undergoing severe criticism. Third (and relatedly), during the formative years of federal hazardous waste policy, the political agenda of the executive branch was actively hostile to Congress's regulatory program. These factors cannot be understood without a brief history of the legislative underpinnings of the government's response.

Waste disposal in America had long been a state and local prerogative, but by the mid-1960s open dumps in urban areas attracted the federal government's attention. In 1965 Congress passed the Solid Waste Disposal Act. This law placed responsibility for waste disposal in the old Department of Health, Education, and Welfare and gave states and localities small grants for improving municipal waste disposal. At the height of environmentalist agitation, Congress revised the 1965 act with the Resource Recovery Act of 1970. This act called for a comprehensive investigation into hazardous waste management practices in the United States, with a formal report to be submitted to Congress. But the entire solid waste management program of the newly created EPA was ignored, according to one early EPA official, as

> an unwanted orphan in an institution which at that time seemed to regard only air and water pollution as legitimate offspring. Few seemed to care, within or outside of EPA. That included the major public interest groups or the environmental community which were still apparently intoxicated by the bold regulatory moves in air and water pollution which engaged the passions and interest of the leaders of the new Agency. That was the way it was. If what the Federal Government does is a reflection of what active public opinion wants—and I believe it usually is—not many people gave a damn about waste, hazardous or otherwise.[12]

By the mid-1970s, the hazardous waste issue, as it became part of a general move to revise the waste recovery law of 1970, began to find some congressional sponsors. Senators Edmund Muskie (D-Maine) and Jennings Randolph (D-Virginia) and Representative Paul Rogers (D-Florida), all important congressional forces in passing the landmark environmental legislation of the early 1970s, began to agitate for new legislation on waste.

Subtitle C of the Resource Conservation and Recovery Act: Entrepreneurial Abandonment

This agitation finally led to the passage of the Resource Conservation and Recovery Act of 1976 (RCRA) and signaled a change in status for hazardous wastes. Subtitle C of RCRA authorized the EPA to develop (1) a definition of hazardous waste;[13] (2) a manifest system to track hazardous waste from its generation to its final disposal; (3) standards for generators and transporters of hazardous waste; (4) permit requirements for facilities that treat, store, or dispose of hazardous waste; and (5) requirements for state hazardous waste programs. Subtitle C was only a small part of the federal government's ongoing pursuit of a national solid waste policy. At the time, the more controversial parts of RCRA involved debate over the extent of the federal government's intrusion into the state and local prerogatives of nonhazardous solid waste management.[14]

The relatively inconspicuous presence of hazardous wastes in Subtitle C of RCRA is all the more striking given the attention that toxic substances in general were receiving in legislation debated and passed simultaneously with RCRA, the Toxic Substances Control Act of 1976 (TSCA). Ironically, the hazardous waste provisions of the RCRA actually benefited from the acquiescence of industry lobbyists: "The industries that ultimately generated toxic wastes—chemicals, petroleum, and mining—were too busy fighting TSCA. They ignored RCRA. Indeed, the CMA [Chemical Manufacturing Association] welcomed the inclusion of hazardous wastes under RCRA. It gave the industry another reason for arguing against the inclusion of waste hazards in TSCA, which was expected to be a much tougher statute."[15]

If regulatory authority for hazardous waste had been housed in TSCA, it would have become a prominent part of pioneering legislation granting the EPA—at the national level of government—wide-ranging and virtually exclusive control over the implementation and enforcement of a federal hazardous substances policy. Instead, placement in RCRA made hazardous waste regulation a seemingly insignificant part of legislation aimed mostly at encouraging states to plan ahead for their larger solid waste problems.

Besides including hazardous waste regulation in RCRA rather than in TSCA, the legislative deliberations over RCRA offer two important elements for understanding the subsequent difficulties in implementing Subtitle C. First, advocates of a federal leadership role in RCRA implementation were defeated at every turn by forces favoring deference to states and localities. Second, the architects of RCRA made a crucial concession to the outgoing Ford administration, whose Office of Management and Budget (OMB) successfully lobbied for provisions permitting HWDFs to continue their existing disposal practices while the EPA hammered out regulations for implementing Subtitle C. This concession let the Ford administration avoid the immediate fiscal impact of rapid implementation of RCRA's hazardous waste provisions, but, more important, it allowed hazardous waste generating and disposal industries to continue status quo ante operations, with no incentive to pressure the EPA for new regulations.[16] The result was unprecedented delay on the part of the EPA in developing regulations to implement RCRA, a process made more complicated by the law's mandate to enforce its hazardous waste provisions, like its other solid waste provisions, primarily through state environmental regulation agencies.

The peculiar status of hazardous waste regulation in RCRA is largely a product of the lukewarm support given it by its sponsors. Senators Randolph and Muskie and Representative Rogers were distracted from their advocacy of a strong federal program by the issues of federalism raised in the solid waste portions of the legislation and by their struggles over TSCA. It should be remembered as well that there was no public outcry at this time to galvanize congressional support behind stronger legislation, and the new Congress that returned from the elections of 1974, even though overwhelmingly Democratic, was no longer uncritically sympathetic to the creation of new environmental programs.[17] Thus, the ingredients of entrepreneurial politics, which might have led to the passage of stronger hazardous waste legislation, failed to materialize.

All these factors conspired to create real difficulties for Subtitle C's implementation. RCRA mandated that the EPA produce hazardous waste regulations by the spring of 1978, but an unmotivated regulatory constituency (the Ford administration's compromise ensured this) and administrative apathy (in its early years, the Carter administration was not supportive of the EPA's role in hazardous waste management) allowed the program to languish. Had the legislation had an entrepreneurial sponsor, committed executive or agency leadership, or an attentive public, the result might have been different. But administrative implementation of social regulation is typically its downfall, and, in the case of Subtitle C, the EPA sat on the legislation until it was too late.

The poignancy of this delay is obvious. If the EPA had developed its regulatory program on time and with quiet deliberation, regulations would have been in place several months before President Carter declared the federal state of emergency at Love Canal—an event that permanently transformed the political climate surrounding hazardous waste regulation. From that point until the regulations were finally promulgated during the spring and summer of 1980 (actually taking effect in November 1980), the EPA weathered intense criticism from Congress and lawsuits from environmentalists for not proposing the regulations quickly enough. Then, when it belatedly proposed the regulations in late 1978 and 1979, the EPA encountered extensive comments and criticism, mostly from industry sources frightened about their liability in the post–Love Canal political climate.[18]

A key component of the EPA's failure to promulgate timely regulations was its essential ignorance of the technical dimensions of the hazardous waste problem. In the most important congressional investigation of the EPA's early efforts under RCRA, the House Subcommittee on Oversight and Investigations issued a stinging denouncement in October 1979, after a year of hearings and investigations, charging that the EPA had, among other things, failed to conduct a comprehensive inventory of the location and contents of hazardous waste disposal sites.[19] In essence, then, the EPA was making and remaking its regulatory proposals in an informational vacuum. Taking a thorough inventory of hazardous waste sites in America would have required years and substantial resource commitments, but public pressure to produce a regulatory program quickly after Love Canal was intense. Built upon a questionable information base, the proposed regulations were particularly vulnerable to criticisms from industry.[20] The atmosphere of controversy threatened the legitimacy of the entire regulatory program.

CERCLA and Superfund: A Child of Crisis

In the midst of the clamor for final RCRA rules for hazardous wastes, the GAO issued a report, dated December 17, 1978, advocating the establishment of a freestanding fund to clean up abandoned hazardous waste sites. The fund was to be financed through fees on hazardous waste disposal. This report, and the EPA's active endorsement of it, were the beginnings of the controversial Superfund law—the Comprehensive Environmental Response, Compensation, and Liability Act (CERCLA) of 1980—a very different piece of regulatory legislation.

If entrepreneurial politics had failed RCRA, such would not be the case with CERCLA. Its passage, its promise, and its problems make CERCLA a textbook case of entrepreneurial social regulatory legislation. It was redistributive in intent; its passage through Congress was a winnowing process ultimately salvaged by dramatic events that further aroused public attention about hazardous wastes; and it eventually became a federal program distorted by the vicissitudes of implementation.

By 1979, EPA officials and key congressmen felt that the biggest hazardous waste problem was not the regulation of operating HWDFs, but cleaning up the closed and, in some cases, abandoned ones. Public concern over hazardous waste at the time was sufficient to inspire environmental advocates in Congress to champion the cause of hazardous waste cleanup. Senators Muskie, John Culver (D-Iowa), and Albert Gore (D-Tennessee) and Representative James Florio (D-New Jersey) became strong advocates of legislation that would finance cleanup of hazardous waste sites through a redistributive tax, primarily on the chemical and petroleum industries. The EPA aggressively supported such legislation and convinced the Carter administration to draft its own bill, and the momentum made strongly redistributive legislation seem certain. But entrepreneurial politics and the fragile legislative coalitions upon which it trades make passage of any such bill problematic.

All versions of the proposed legislation were novel in several regards. First, they were to enable the EPA to respond swiftly to cleaning up hazardous waste sites around America without becoming endlessly entangled in regulatory hearings and litigation. As one EPA official put it at the time, CERCLA was to call forth "shovels first, and lawyers later."[21] Second, the legislation was strongly redistributive. Industries associated with the generation of hazardous wastes were to pay in advance into the Superfund to finance both remedial and emergency cleanups, and Congress would add a relatively small appropriation from general revenues. When sites were identified for cleanup, the EPA was to determine the parties responsible for their contamination and then try to force those parties to clean up the sites. If complications delayed cleanup, however, the EPA was to begin cleanup immediately by tapping the Superfund, then hold the responsible parties liable to repay the Superfund for the cost of the cleanup.

The redistributive impact of the legislation caused enormous controversy. The chemical and petroleum industries lobbied hard against the funding scheme, arguing that the Superfund should be financed entirely from general revenues. After all, industry sources argued, hazardous waste was "everybody's problem."[22] Further, the issue and extent of responsible party liability inspired vigorous debate, as did various schemes for victim compensation. Finally, legislators disagreed over the role the states should play as partners with the EPA in the administration of the cleanup program.

Congressional attention to CERCLA waxed and waned in 1979, and it was not until 1980 that earnest movement on the legislation began.[23] Such mercurial attention to an issue is characteristic of the entrepreneurial politics of social regulation in general. Congress became occupied with other environmental matters, particularly in response to the energy crisis. The lapse of attention to CERCLA proved crucial as opponents of the legislation gained time to mount a counteroffensive. Further, the legislative coalition supporting a strongly redistributive CERCLA began to decompose. Senator Muskie became Carter's secretary of state in April 1980, and Senator Culver was forced to fight for his political life in a reelection bid he would ultimately lose.[24] The removal and distraction of CERCLA's strongest Senate advocates cast doubt upon the future of the legislation. Then, just one week after Muskie's appointment, the incident described in the opening of this book occurred: chemical fires at the port of Elizabeth, New Jersey, raged out of control on the doorstep of New York City. In response, the House's version of CERCLA, championed by New Jersey's Florio, began to move toward passage. Meanwhile, the Muskie-Culver version, a much stronger bill in terms of the size of

Superfund it envisioned and in terms of its liability requirements for industry, began to lose steam.

Opponents of the legislation were able to stall its consideration until after the presidential election of 1980, which swept Ronald Reagan and a Republican Senate majority into office. Congressional Democrats and the Carter administration were forced to seek final passage of the legislation in the lame duck legislative session that followed the election. The CERCLA that was signed into law in December 1980 turned out to be a desperate compromise, one in which key features were jettisoned to salvage a workable program before time ran out. Significantly, the legislation, as passed, retained its redistributive design and its federally oriented character, with the EPA maintaining a dominant position vis-à-vis the states in the administration of CERCLA.[25]

Stung by congressional, environmentalist, and industry criticism for its slow progress on RCRA regulations, the EPA sought to redeem itself by taking the "emergency response" character of CERCLA seriously and planning for its swift implementation well before the bill was finally passed:

> The preimplementation project allowed [the EPA's] Superfund organization to develop its area of distinctive competence before operational pressures began. It enabled EPA to minimize the effects of mistakes always made by new organizations. In addition, the Superfund organization was able to preempt turf battles over Superfund by working in the area at a time when risks were high (the absence of legislation) and rewards for involvement were low (due to the absence of significant new resources to participate in Superfund). . . . Recruitment of staff was well underway. Year-long contract procurement processes had already begun. Internal regulation review processes had been underway for several months.[26]

Office morale in EPA was high when President Carter, in one of the last official acts of his presidency, signed the executive order authorizing the EPA to implement CERCLA.[27]

Superfund Implementation: The Design of Failure

The political context within which CERCLA developed was very different from that of RCRA.[28] CERCLA was quite clear in its mandate, and, as mentioned above, the EPA staff charged with its implementation were, at least initially, strongly motivated. This, of course, jeopardizes one of our arguments about the new social regulation— that implementation is its downfall for lack of entrepreneurial politics.

But the supportive bureaucratic context surrounding CERCLA's early implementation became hostile in 1981, as the Reagan administration actively challenged the assumptions under which CERCLA was passed. David Stockman,

President Reagan's new director of the Office of Management and Budget, led a group of Reagan officials who reinterpreted CERCLA's mandate for federal action, charging the EPA to delegate its Superfund powers to the states, a move consistent with the Reagan administration's New Federalism. As a congressman, Stockman had bitterly opposed the concept of a Superfund cleanup managed by the federal government.[29] This turnabout apparently caught the EPA off guard and resulted, once again, in a poorly planned implementation of federal hazardous waste legislation.[30]

CERCLA had begun as an issue involving citizen pressure upon government to respond to a social crisis. It now became, in implementation, an internecine partisan struggle between branches of government over the redistributive impact of social regulatory policy. The seeds of public distrust, sown in the delays and controversy surrounding the promulgation of RCRA regulations, were nurtured during the early years of the Reagan administration. The pattern of interaction between Congress and the EPA from 1981 to 1983 resulted in the delegitimation of federal hazardous waste policy, just when the general public's concern about the problem (as we shall discuss in chapters 6 and 7) was becoming more and more acute and well informed.

In spite of public opinion, which continued to support federal environmental protection,[31] the Reagan administration appointed Anne Gorsuch, an avowed opponent of hazardous waste regulation, to be the administrator of the EPA.[32] Testifying in July 1981 before the U.S. House Subcommittee on Commerce, Transportation, and Tourism, Gorsuch asserted that she had made "the full implementation of Superfund the highest priority of the Environmental Protection Agency." Members of the subcommittee, which was chaired by James Florio, listened skeptically. With the passage of the Senate majority into Republican hands, Florio would use his subcommittee chairmanship over the next few years as a platform both for attacking the EPA's handling of RCRA and CERCLA and for strengthening the legislation surrounding hazardous substances regulation in general.

The continuing struggle between Gorsuch and Florio revealed grave discontinuities in the government's endeavors at hazardous waste regulation. The EPA's top administrators consistently distributed bland assurances that, with the states' active cooperation in RCRA programs, the hazardous waste problem was under control and that Superfund cleanups would largely "pay for themselves" through enforcement actions against responsible parties and through state participation in cleanup supervision.[33]

Setting the tone for the EPA's general approach to regulation, Gorsuch strongly advocated the need for "better science" in the EPA, and she initiated much-publicized programs to accomplish her avowed goal. Such a proclamation appeals to the managerial rationale for legitimating policy, based as it is upon the myths of science as a guide for policy-making. In fact, Gorsuch used the better science campaign to mask a political agenda. Between 1981 and 1984, she actually cut the EPA's research budget by 50 percent, leading one analyst to claim,

> In light of EPA's serious knowledge gaps in rapidly developing areas such as toxic chemicals, along with the agency's need to reassess early standards set on weak evidence, a serious commitment to improving EPA's science would have dictated simply a reallocation rather than a reduction of EPA's research budget. A more plausible explanation of Gorsuch's policy, therefore, is that her primary goal was not improving science, except where convenient and free of cost, but implementing the Reagan/Stockman policy of deregulation and domestic budget cuts.[34]

The agency's activity in regard to hazardous waste also contrasted with its rhetoric. Gorsuch asked for deep budget cuts in its RCRA program (which was reduced from $150 million in fiscal year 1981 to $87 million in fiscal year 1983), and the EPA consistently temporized in identifying Superfund sites and in authorizing expenditures for cleanups. The agency suspended its regulations on surface impoundments and incinerators and announced its intention to cut liability insurance requirements for HWDF operators and to allow existing operators to expand operations by 50 percent without prior EPA approval.[35]

A managerial approach symbolized by better science rendered concessions to democratic visions of regulation unnecessary. Thus, Gorsuch dismantled EPA's citizen participation programs, eliminating the Office of Public Awareness, reducing memberships on advisory boards, and shortening public comment periods on agency proposals.[36] By ignoring public input, the EPA could better pursue its own idea of an "objective public interest," according to the logic of the managerial perspective. In fact, the lack of public visibility enabled the agency, under Gorsuch, to pursue policies that would soon become controversial.

Rumors circulated that the EPA's efforts to force quick settlements with responsible parties were actually sweetheart deals that allowed the parties to buy out of their continuing liability for hazardous waste sites while paying for only superficial cleanup. Charges also circulated that Superfund moneys were being withheld from congressional districts in which Democrats might stand to gain from Superfund activity.[37] At the same time, programs implemented under RCRA and CERCLA were already revealing that earlier estimates of the hazardous waste problem were simply the tip of the iceberg. Also, new research produced by the GAO and the Office of Technology Assessment, largely at the behest of Florio, was documenting not only the enormous size of the problem, but also the potentially staggering costs involved in cleaning up the sites estimated to exist. Further, these reports consistently criticized the EPA for not meeting the growing challenge of hazardous waste regulation and gave Democrats in Congress a continuous supply of ammunition with which to challenge the EPA's credibility.

Public and congressional protests peaked in February 1982, when the EPA suspended the RCRA rules that banned the land-filling of containerized liquid waste. The rule was reinstated in May 1982, but the credibility of the agency in the area of hazardous waste management was severely shaken. Bipartisan congressional attention was aroused and several investigations ensued. Still, the EPA saw nothing but success in its efforts, praising the private sector for its initiatives in site cleanup and hazardous waste generators for "assuming their responsibilities" and for their willingness to settle.[38] Congressional pressure was taking its toll, however, and when Gorsuch claimed executive privilege and resisted a subpoena to turn over EPA's Superfund expenditure documents to the House Subcommittee on Investigations and Oversight in December 1982, she was cited for contempt of Congress. In February 1983, Rita Lavelle, EPA's chief enforcement official for Superfund cleanup, under suspicion of offering perjured testimony to Congress, was fired. A month later, Gorsuch herself[39] resigned and was followed out by most of the political appointees brought into the EPA under the Reagan administration.

While political scandal surrounded EPA, information about the hazardous waste problem was accumulating. CERCLA required the development of the Emergency and Remedial Response Information System (ERRIS) to inventory and prioritize hazardous waste sites across the country for inclusion on a so-called National Priority List (NPL) of sites as part of the National Contingency Plan for hazardous waste cleanup. The NPL began with 115 Superfund sites and gradually grew to a "final" list of more than 400 sites in early 1983, its growth precisely coinciding with the EPA's deepening political scandal. The expansion of the information base, rather than providing the EPA with more certain and more credible evidence to legitimate, in managerial terms, its policy pronouncements, actually became the chief indictment of the agency's scandalously ineffective administration. Combining two prime ingredients for entrepreneurial politics—scandal and disaster—the public learned about hazardous waste and its regulation primarily from media reports that focused on potential dangers to the public and on the EPA's mismanagement of hazardous waste policies.

The extent of the problem and the EPA's failure to deal with it were driven home by the revelations of the NPL. These revelations came not in the abstract but in very real terms, in states and communities in which hazardous waste sites were discovered and in which Superfund status made local news, thus radically affecting local populations and mobilizing their concern. The delegitimation of the government's regulation of hazardous waste had a tangible local impact, awakening citizens both to the dangers of hazardous waste and to the ineffectiveness of government actions addressing the problem. These became crucial components of the distrust that citizens and communities manifested in our case studies of hazardous waste regulation in the states. Their distrust led, in part, to rejection of government bids to site HWDFs at the local level and to forging their own grassroots initiatives.

The replacement of Anne Gorsuch with William Ruckelshaus eased tensions about the EPA's future in a general way, but its hazardous waste programs continued to come under attack. First, there was the problem of getting the job done. Between 1981 and 1983, as estimates of the sites eligible for Superfund cleanup soared to between 1,400 and 2,200, the EPA completed just 2 remedial, or long-term, cleanups, while only 23 were in the planning stages.[40] In March 1984, Hugh Wessinger, a GAO official, reported to Representative Florio's subcommittee that the cost of Superfund cleanups might rise as high as $10.5 million per site for groundwater contamination alone. This estimate exceeded earlier EPA assurances that the total cleanup cost per site would be in the $4 million range. In effect, then, the EPA had simply covered up its unwillingness to accomplish Superfund cleanups with facts and figures that diverted public and congressional attention away from the reality that the job was much bigger than at first anticipated and that it was simply not getting done.

In a comment that proved ominous for the prospect of redistributive regulation of hazardous wastes at the local level, the Wessinger report noted that state cleanup funds which taxed waste generators were consistently falling short of projected revenues. As we shall see in the empirical analysis that concludes the chapter, this finding was indicative of a pattern of failure among the states to impose the redistributive burdens of environmental regulation on offending industries.

Meanwhile, the NPL, which was to be updated every year, continued to grow, while few cleaned-up sites were removed from the list. Now controversy surrounded the criteria for listing sites and the timing of the release of candidate sites for inclusion on the NPL. Representative Florio produced an "overdue list" of 203 candidate sites in September 1984, leaked to him by a frustrated EPA staff member. He claimed that release of the list was stalled to prevent its having a negative impact on the upcoming elections and to diffuse support for CERCLA amendments being deliberated by Congress at the time:[41] "The Reagan administration is afraid that if the hundreds of thousands of Americans who live around these toxic waste hazards find out that they have been designated as Superfund sites, the public outcry will propel this legislation into law."[42]

Within days, the EPA added 208 sites to the NPL, blaming the delay on the lengthy site selection process. Immediately following the elections of November 1984, President Reagan reluctantly signed RCRA reauthorization into law (as the Hazardous and Solid Waste Amendments of 1984), which toughened RCRA regulations and imposed deadlines upon the EPA.[43] But the White House and Republican leaders in the Senate were able to forestall passage of legislation to extend CERCLA for nearly two years, leaving the program to limp along on a maintenance budget while the NPL grew to nearly 850 listed and proposed sites.[44]

The problem of scientific and technical uncertainty, which had done so much to discredit the EPA during the early years of the Reagan administration, surfaced again in 1985. Congressional and environmental critics blamed the Superfund's site selection process for more than just delays, claiming that its "hazard ranking system" was unscientific, arbitrary, influenced by politics, and biased against environmental (as opposed to human health) effects.[45] These criticisms forced reconsideration of the hazard ranking system just before passage of the Superfund Amendments and Reauthorization Act (SARA) in October 1986.

The SARA legislation reauthorized Superfund for five years and substantially increased its funding limit to $9 billion (including $500 million for cleaning up leaking underground storage tanks). Reflecting the federal government's ability to adopt policies that are difficult to pursue at other levels of government, the funding retained its originally redistributive balance, with the petroleum industry shouldering most of the new tax burden and the chemical industry bearing relatively less, when compared to the revenue balance in the original legislation of 1980. Not coincidentally, SARA, like CERCLA before it, benefited from media coverage of yet another toxic crisis: in December 1984, an industrial accident in Bhopal, India, involving a chemical explosion at a Union Carbide plant, left 2,000 dead and 100,000 wounded. The news hastened congressional consideration of the SARA legislation and encouraged Congress to include a significant freestanding title (Title III) authorizing community right to know, that is, access to information on the manufacture, use, and storage of chemicals at local industrial facilities, primarily for emergency planning purposes.

In spite of the Bhopal incident, overwhelming and veto-proof majorities behind SARA in both houses of Congress, and support for the legislation within EPA, implementation of SARA by the agency brought charges of foot-dragging. A GAO report in 1988 suggested that an astronomical 425,380 potential Superfund sites existed in the United States, of which nearly one-quarter had a high likelihood of being contaminated with hazardous waste. The report criticized the EPA for placing too little emphasis upon detection, inventory, and assessment of new sites, leaving these tasks up to the states, which, the report claimed, were provided little guidance and less money from the EPA to do the job.[46]

Although the agency may have been emphasizing assessment and cleanup of already existing Superfund sites, the results certainly did not indicate much progress there either. As it began to implement SARA, the EPA itself noted that only fourteen of the NPL sites had been permanently cleaned up in the six years since CERCLA was originally implemented, while its estimation of potential Superfund sites had grown from 2,200 to 4,000 sites.[47] The Office of Technology Assessment issued a study in June 1988 that updated its long-standing criticism of the agency's implementation of CERCLA.[48] In response to the criticism, the EPA continued to insist that progress was being made.[49] Thus, the pattern of attack and denial begun in 1981 continued to the end of the decade.

As the Bush administration took over in January 1989, pledging an "environmental presidency," the EPA, under the administration of William K. Reilly, continued to accumulate Superfund sites on the NPL (the total as of April 1989 stood at roughly 1,200 sites), and the public continued to rank active and abandoned hazardous waste sites as the most serious environmental risk facing citizens in the United States in 1989. Ironically, an internal poll of EPA staff in 1988 ranked the hazardous waste issue in the low to moderate category, while the "greenhouse effect," in contrast to the public's relatively low ranking of that issue, was given the highest ranking.[50]

The disjuncture between public attitudes toward the hazardous waste problem and the EPA's official response makes a fitting summary to the events described here. It indicates the dominant managerial orientation of the EPA and its inability (and perhaps unwillingness) to address citizen concerns over the issue of hazardous waste in their communities.[51] The disparate perceptions of the agency and its public are a telling indication of the lack of policy dialogue over the issue of toxics regulation.

Regulatory Reform, the New Federalism, and State Regulation

After a decade of implementation, the EPA's handling of RCRA and CERCLA produced the definitive example of how agencies can defeat even the most adamantly expressed legislative intent and public desire for action. By replacing the legislative efforts of political entrepreneurs with the hidden agendas of "interest group liberalism," the Reagan administration's actions confirm our assertions about the travails of social regulation once it enters the realm of bureaucratic politics. But the EPA and the Reagan administration cannot be held totally accountable for the failures of federal hazardous waste regulation. As an integral part of the new social regulation, hazardous waste regulation was under a much broader attack during the years of its development and implementation, an attack articulated in our discussion of the various waves of reform that have influenced federal regulatory policy during this century (see chapter 2).

The Languages of Reform and Hazardous Waste Regulation

The ideology of "regulatory reform" in the late 1970s and early 1980s placed a dual emphasis upon economic efficiency and decentralization, shifting the locus of regulation away from the federal government and toward the states. In effect, reformers were blending the conventional languages of regulatory debate and judging regulation, particularly the new social regulation, by standards that combined the supposedly objective, managerial assessment of cost-benefit analysis, on the one hand, with a call for more pluralistic regulatory initiatives from the state and local levels, on the other. We argue that hazardous waste regulation has been both a product and a victim of this episode in regulatory reform as it has evolved within the structure of federalism.

Regulatory reform's managerial emphasis on efficiency in the late 1970s was largely inspired by trenchant conservative criticism and research aimed at demonstrating the tremendous unproductive cost—compared to the social benefits—of the new social regulation on business.[52] Of course, the new social regulation was supposed to cost something, given its redistributive intent, and its largely noneconomic benefits were hard to capture in dollar terms.[53] But that did not deter the advocates of efficient regulation. The managerial language found a new voice among conservatives, who expropriated some of the old Progressive rhetoric—but to different ends—in arguing that the public interest in regulation was simply a matter of economists' calculations.

Cost-benefit analysis of regulation—or, in the phrase used at that time, making regulation pay for itself—became a particularly attractive theme during the tough economic times of the late 1970s. Both Presidents Ford and Carter issued executive orders[54] that required agencies to assess the costs and benefits of proposed regulations before issuing them. This regulatory theme was later borrowed by presidential candidate Reagan and strengthened following his election in 1980 in Executive Order 12,291, which required agencies' proposed regulations to pass an OMB cost-benefit analysis before being approved. The managerial mandate of cost-benefit analysis placed an intangible burden of proof upon regulators in Washington to demonstrate that their programs would produce results more desirable than those of unfettered market operation. Hazardous waste regulation, challenged with making costly redistributive policy on the basis of uncertain scientific and technical information, was thus launched into hostile regulatory waters.

At the same time, the call for decentralization of the new social regulation, which culminated in Reagan's New Federalism, was framed within pluralist discourse, projecting the states as fifty independent laboratories, all pursuing experiments in regulation, closer to the local level and more capable of properly balancing the diverse interests of those being regulated. Certainly RCRA was infused with this new approach to regulation. It allowed states much greater flexibility in complying with the EPA's mandates than earlier environmental programs had. The parts of RCRA that applied to solid waste management allowed great state autonomy, in part out of respect for the traditional state and local leadership in waste management issues, but also in response to reformers' arguments that the states could do a better job than the federal government if given the authority.

Because it broke new regulatory ground, Subtitle C retained a modicum of centralized control, but the deference given to the states in the other parts of RCRA often carried over to the interpretation of its hazardous waste provisions during EPA implementation.[55] In effect, the formative stages of hazardous waste regulation were caught in an organizational irony: the centralizing and redistributive nature of a national system of hazardous waste regulation was contained within a decentralizing and allocational vehicle of a solid waste management program designed for state-level implementation.

Such was the context of regulation when CERCLA was proposed. On one side, the public was clamoring for a solution to the Love Canals dotting the American landscape; on the other was a moribund economy and policymakers who were convinced that regulation in general was a burden the economy could hardly bear. Further, the presidential elections of 1980 loomed, with the promise of a so-called regulatory revolution to be led by Reagan's presidency. Yet the discontinuities between regulatory languages allowed policymakers to avoid confronting these contradictory demands. So, CERCLA survived passage as federally oriented and redistributive legislation, primarily because of the urgency of the hazardous waste problem and the persistent efforts of political entrepreneurs. But its thrust was blunted by the Reagan administration's skillful exploitation of reform sentiments, as expressed in the use of, first, managerial and then pluralist languages aimed at legitimating interpretations of the law that stifled its redistributive impact and resulted in little progress toward cleaning up hazardous waste sites.

On the one hand, the implementation of CERCLA was turned over to administrators who, despite justifying their behavior primarily in managerial terms by citing better science, were effectively constructing a smoke screen of technical information aimed at precluding public inquiry into the progress of Superfund cleanups. They were in fact protecting the powerful interests of industry from having to absorb the costs of regulation, while keeping those interests far from the reach of citizens who were paying the price of living near potentially dangerous hazardous waste sites. The language of regulatory reform enabled administrators to disguise their intent within the managerial language. But the product, after eight years of the Reagan administration, was not regulatory reform but regulatory relief: "They perceived their mission simply in terms of reining in the professional bureaucrats whom they saw as undermining the political economic well-being of the United States. Despite their initial claims about introducing scientific criteria and cost-benefit analysis, their actions bespoke an attempt merely to 'ratchet down' regulation."[56]

On the other hand, the Reagan administration also justified its alleged reform of regulation in pluralist terms by embracing the theme of regulatory decentralization. The New Federalism became an argument for returning regulatory power to the states. In our view, however, the uses of pluralist rhetoric disguised a bid to ensure that a broad range of federal regulation would be defeated by the structural incapacity of the states to implement such regulation. The rhetoric of reform used by David Stockman, Anne Gorsuch, and Rita Lavelle consistently minimized the need for an active federal role in the administration of RCRA and CERCLA. They constantly repeated themes of deregulation (championing the private sector's sense of responsibility in cleanups) and delegation to the states (extolling the effectiveness of the localized regulation that state governments could allegedly supply).

Such rhetoric disguised sweetheart settlements with hazardous waste dumpers that landed Rita Lavelle in prison. It also justified reducing the EPA's RCRA enforcement budgets through arguments that the states could more effectively assume the responsibilities of environmental protection.[57] The result was to diffuse through the system of federalism the redistributive burden of hazardous waste management and cleanup, a burden that Congress had intended to be borne by the industries responsible for the problem.

Under the assumptions of the New Federalism, the states became the central focus of the federal government's efforts to regulate hazardous waste. We argued above that the successful implementation of redistributive regulation in a federal system will face severe structural constraints precisely because of the states' vulnerable position vis-à-vis redistributive policies (see chapter 4). It would be hard to prove that Stockman and others who used the language of the Reagan administration's New Federalism knew in advance that the redistributive impact of hazardous waste regulation would be difficult for the states to implement.[58] But we can assess empirically the structure of state environmental regulation at the time such arguments were being made. Were the states good candidates for shouldering the redistributive programs of hazardous waste regulation?

State Environmental Regulation: An Empirical Analysis

The Reagan administration's assertions that the states were better able than Washington to regulate hazardous waste were presumably based upon evidence of such accomplishments. These assertions contrast sharply with our arguments in chapter 4 about the structural limitations of states in a federal system in implementing redistributive policies. Here, we investigate empirically these competing conceptions of state regulatory capacity.

Our focus is on the performance of states in the area of environmental regulation generally. We are investigating whether or not state environmental programs were sufficiently evolved by 1980 to assume the redistributive task of regulating hazardous wastes. This would require, first, a developed regulatory infrastructure, one sufficiently funded and staffed to accept the challenge of enforcement. Second, the agencies charged with implementing federal hazardous waste policies would require state legislation to give them enforcement powers consistent with their federal mandate. Third, to address effectively the problem of abandoned hazardous waste sites, states would need the funds necessary to match Superfund requirements.

If state environmental programs had a track record of implementing redistributive environmental policies in general and specifically those affecting the industries targeted for hazardous waste regulation, then the assumptions of the New Federalism would have been well founded. If, on the other hand, programs failed to finance themselves redistributively, then they were not likely candidates to carry forward the legislative intent of RCRA and CERCLA. Instead, the New Federalism would simply spare industry the redistributive burdens of federal hazardous waste regulation by diffusing its impact at the state level, either through ineffective enforcement or by covering the costs of regulation with allocations from general revenue funds.

Of course, many interrelated factors affect the capacity of state environmental programs to impose redistributive policies, and we have taken these into account. A straightforward way of considering the capacity of state environmental programs at the time arguments for the New Federalism were being articulated is to analyze and compare state expenditures on environmental regulation and private industry expenditures on land and water pollution abatement for this period. To this end, we identified public and private expenditures on pollution control as dependent variables in separate sets of regression equations. The specification of these and the independent variables included in our analysis is summarized below. A complete account of our analysis, including detailed descriptions of dependent and independent variables and further discussion of the propositions we tested, as well as tabular presentation of the intercorrelation matrix and regression results from our analysis are presented in the Appendix. Here, we simply outline our research design, discuss our analysis generally, and report a summary of our findings.

Propositions and Data: A Summary During 1980–81, the crucial time frame for determining the regulatory capacity of the states on the eve of the New Federalism, there were no data detailing specifically state-by-state public and private expenditures on hazardous waste regulation. Instead, we used data from fiscal year 1980 covering general expenditures on land and water quality control by both the public and private sectors in each state as a basis for determining the extent of redistributive environmental regulation in the states.[59] These assumptions and our conclusions have been supported in subsequent research by others.[60]

We ran two sets of regression equations. One used a dependent variable representing total public expenditures in each state on land and water regulation during 1980, and the other used a dependent variable reflecting the expenditures for this purpose by those private industries producing the most hazardous waste, relative to the size of those industries in each state. These industries would be the ones targeted for the redistributive burdens of regulation in subsequent years. If states were directing the burden of general environmental quality control expenditures to these industries in 1980, then there would be a strong case—consistent with the New Federalism—for their ability to redistribute hazardous waste costs to these industries in the future.

Our summaries of each independent variable used in the regressions follow here and include its predicted effect based upon the assumptions of the New Federalism contrasted with the predictions of our model of regulation in a federal system (as set out in chapter 4):

1. *The state's need for hazardous waste regulation—the relative severity of each state's hazardous waste problem.* The New Federalism assumes that the objective size of a state's hazardous waste problem dictates the imposition of a regulatory burden upon the industries producing it. Our model suggests that neither public nor private expenditures in a state will be related to the extent of the problem; rather, the state will be limited, whatever the level of need, by the size of its already occurring general revenues and budget.

2. *The state's fiscal capacity—the size of each state's budget for fiscal year 1980.* This should be independent of state and private expenditures on environmental regulation, according to the New Federalism, but because our theory of regulatory federalism suggests that states will consistently transform redistributive policies into allocational ones drawing from general revenues, the size of the state's budget should have a strong influence on that state's pollution expenditures and little or no influence on private expenditures.

3. *The quality of state legislation.* If states were ready to pursue federal hazardous waste mandates in 1980, the quality of their laws in that area should be a predictor of public and private expenditures on environmental regulation. But our arguments suggest that strong laws are often symbolic and do not necessarily translate into increased expenditures in either sector.

4. *Industry's strength—the regulated industries' contributions to the state's economy and employment of its workforce.* The New Federalism predicts no relation here, but we expect a negative one between influence and expenditures in both sectors. If industry and the workers it employs contribute to a

state's economy, we expect that the state will be reluctant to impose regulatory burdens because of vulnerability to "economic blackmail."

5. *Industry structure—whether the targets of environmental regulation in the state are large, oligopolistic firms or small, competitive firms.* Large firms may, in fact, welcome some forms of environmental regulation: for public relations purposes, to protect long-term investment in pollution equipment, to prevent future liability, and even as a means of limiting entrants into their market. An economy composed of smaller, competitive firms should resist regulation as a relatively large part of their marginal costs of doing business. Again, the New Federalism assumes no connection between industry structure and public or private pollution expenditures.

6. *Public interest group strength—the significance of each state's environmental lobby.* The New Federalism suggests that states can better balance interests in environmental regulation than the federal government can and, thus, would predict that redistributive private expenditures on pollution abatement would vary directly with the strength of a state's environmental movement. Our model suggests that state governments translate environmentalist pressures into larger allocational commitments to environmental regulation from general revenue funds, thus avoiding the imposition of redistributive costs on industry.

7. *The state's partisan structure—the dominant political ideology or existence of electoral competition in a state.* Again, if the states are more responsive to public preferences for environmental regulation, as the New Federalism implies, then a state dominated by proregulation Democrats or a state with a competitive party structure eager to appeal to voter concerns about the environment should require more public and private expenditures on pollution abatement. In contrast, we argue that redistributive issues transcend partisan ideology at the state level. Our model suggests that neither party can afford to offend industry with a redistributive regulatory burden, and there will be no relation between partisan structure and environmental expenditures.

A Summary of Findings Reference to tables 1 and 2 and to the accompanying explanation in the Appendix indicates that our analysis was both internally consistent and explained a substantial portion of the variance in public and private sector spending on land and water pollution abatement. More important for the assumptions of the New Federalism was the pattern of relations among the variables. With only one minor exception, the relations predicted in chapter 4 prevailed over those implicit in the New Federalism. Consistent with the tendency of state governments to transform redistributive mandates into allocational policies, the primary determinant of state environmental expenditures in 1980 was simply the size of the state's budget.

Our analysis also showed that states often passed strong legislation in response to the severity of their hazardous waste problem, but subsequent funding of related programs bore no relation to either the extent of the problem or the quality of the legislation. Regulated industries had a strong negative influence on the redistributive impact of state regulatory efforts, especially if they were critical to the state's economy or if their composition made them structurally vulnerable to the costs of regulation. Neither the ideological predisposition nor the threat of electoral defeat seemed to affect policymakers' attraction to the general revenue fund as the allocational answer to the states' environmental problems. The political clout of environmental groups reinforced rather than reversed the allocational tendencies of state policymakers.

The only exception to our predictions was the positive relation between the size of the labor force in the affected industries and state expenditures on environmental regulation. Nevertheless, the relation between labor and private expenditures was negative, as we predicted. Perhaps workers caught in the cross-pressure of environmental concern and economic self-preservation resolved their dilemma by encouraging policymakers to draw upon public funds for environmental regulation instead of saddling industry with costs that might threaten wages or jobs. In effect, then, labor, like the environmentalists, appears to have influenced the tendency of state policymakers to use public rather than private funds to pay for environmental regulation. Advocacy for state environmental regulation was, in 1980, just another interest in the pluralistic competition for shares of state budgets. In spite of the limited funds available to state governments for meeting their other responsibilities, they were consistently unable to impose the burdens of environmental regulation upon industry.

Hazardous Waste Regulation in Three States: An In-Depth Analysis

So far, our analysis leaves open the question of exactly how individual states responded to the challenge of regulatory responsibility after 1980. Our quantitative analysis sets the stage for investigating the readiness of states to implement the specific mandates of RCRA and CERCLA. We focused our analysis on the connection between the competing languages of regulatory debate and policy outcomes in three states—New Jersey, Ohio, and Florida.

In the rest of this chapter, we describe the events shaping each state's recognition of its hazardous waste problems, the relevant regulatory infrastructure (including federal-state relations), legislation, and funding as well as the key political actors in the area. In addition, we address the politicoeconomic structure of each state, thus fleshing out in qualitative terms the variables used in our quantitative analysis. With this as background, in chapter 6 we analyze the political struggles over HWDF siting in the three states in order to understand how the competing languages of regulatory discourse are used by policymakers, citizens, and businesspeople.

Three case studies, of course, cannot fully reflect the diversity of policymaking in all fifty states; so we chose states that faced severe hazardous waste problems during the 1980s and that varied along politicoeconomic dimensions likely to affect social regulation. All three states ranked high in the overall production and use of hazardous substances, and all placed a large number of sites on the EPA's Superfund NPL. In all three states, hazardous waste regulation was a troubling public issue, particularly during the decade of the 1980s, when we conducted our investigation.[61]

Beyond this common policy problem, the three states varied markedly in their political, economic, and social characteristics. Throughout our study, New Jersey was a state in transition. When we first visited, it was in serious economic decline. Indeed, the threat of industrial flight was a central dynamic in hazardous waste regulation. By the end of our study, however, the state had restructured much of its economic base and had benefited from the booming "bi-coastal" economy of the later Reagan presidency. These changing economic conditions altered the balance of power among environmentalists and industry lobbyists in the struggles over regulation. In contrast, Ohio was a Midwestern rustbelt state struggling with a declining industrial base during the 1980s. Politically, the state balanced a conservative Republican, small town political culture against an urban Democratic party rooted in a progressive, labor union heritage. Florida was a sunbelt state grappling with rapid population growth and an expanding but increasingly fragile economy based upon tourism, international finance, real estate, and export agriculture. Along with this economic admixture, the state retained a conservative political culture suspicious of its small, hard-pressed public sector.

New Jersey

Because New Jersey has one of the most severe hazardous waste problems in the nation and has struggled with the problem longer than any other state, it was an obvious choice for our study.[62] The state was the cradle of the chemical, pharmaceutical, and petroleum industries in the United States, beginning in 1840, when William Colgate moved his soap plant from New York to Jersey City. By the late 1800s, Bayonne was a center for refining crude oil pumped from wells in Pennsylvania. Agricultural chemical enterprises grew to support the large truck farming industry that gave the Garden State its nickname. All of these industrial developments generated wastes, some of which were hazardous. However, the bulk of the state's hazardous waste problem resulted from the phenomenal expansion of the petrochemical market after World War II. As our study began, New Jersey had the second largest chemical industry in the nation, employing 95,000 people in more than 120 plants and 1,000 research and development facilities. Historically, the safe handling of this industry's hazardous wastes was generally ignored, leaving an enormous generation, disposal, and cleanup problem in a state with the highest population density in America. To make matters worse, organized crime, long dominant in the waste hauling business, was reported to be moving into hazardous waste disposal in the state, making regulation even more difficult.

As hazardous waste became an object of public concern, systematic investigations revealed the scope of the state's problem. By the end of our study, there were about 1,200 sites in the state listed by the New Jersey Department of Environmental Protection (DEP). The very worst site on the EPA's NPL was the Lipari Landfill in Mantua, New Jersey. There were 2,500 firms registered as producers of at least one hundred kilograms of waste per month; and most were concentrated in the densely populated northeastern corner of the state. However, no region of the state was immune: every county listed at least 9 major producers of hazardous waste. All of these threatened a vulnerable groundwater supply: 3.5 million people, or 45 percent of the state's residents, drew their drinking water from aquifers protected by only a thin covering of highly permeable, sandy soils.

The state's location in the crowded northeastern industrial corridor also means that a wide variety of hazardous substances have been shipped through the state. In 1983, as documented under RCRA, more than 12 million tons of hazardous wastes from in-state and out-of-state generators moved through New Jersey. In that year alone, some 28 commercial disposal facilities received an estimated 400,000 tons of hazardous wastes. In retrospect, it seems probable that most hazardous wastes were disposed of in unsafe ways. For example, a DEP study in 1979 found that 3.2 million tons of the 3.9 million generated in the state were dumped in the ocean.[63]

Aside from the fact that New Jersey was one of the first states to confront the hazardous waste problem, several features of its political economy made it attractive as a case study. The state possesses an enormous and politically potent petrochemical industry, making redistributive hazardous waste regulation potentially difficult to implement. But during our study the state managed to reduce its dependence on the old industrial infrastructure by restructuring its economic base.

The restructuring accompanied a stunning economic revival. For the entire period from 1980 to 1984, New Jersey's gross state product (GSP) grew just 10.5 percent; but in the next year alone, its GSP grew 8.0 percent, and in the next,

8.8 percent. Increases in New Jersey's GSP outperformed those of the gross national product (GNP) between 1985 and 1986 by 3.0 percent. The unemployment rate in New Jersey actually fell between 1980 and 1984 (from 7.2 percent to 6.2 percent) and continued to fall through 1987 (to an even 4.0 percent). Finally, between 1980 and 1987, New Jersey moved from being the fourth to the second wealthiest state in the country (in per capita income), behind only Connecticut.[64] This recovery enabled policymakers to raise the substantial funds required to implement the state's innovative regulatory legislation.

But the recovery was uneven within the state, and the old manufacturing areas (for example, Newark and Trenton) remained depressed. As we shall see, the recovery's unevenness introduced another sort of economic dimension into the issue of hazardous waste regulation. More than in the other states we visited, local citizens' groups have detected injustice in the state's hazardous waste decisions and have articulated a strongly communitarian challenge to the arguments of state officials, attacking even the statewide environmental lobby on HWDF siting (a point we revisit in chapter 6).

Finally, New Jersey boasts of an interesting cast of political figures whose careers have been shaped by the hazardous waste issue. A moderate Republican, Gov. Thomas Kean, took credit for the state's remarkable economic recovery. At the same time, he projected an image of being extremely tough on environmental pollution in general and on hazardous wastes in particular. But when Kean completed his two terms as governor, the state's peculiar off-year gubernatorial election in November 1989 pitted Democratic Congressman James Florio against a relative newcomer to statewide politics, Congressman Jim Courter. Florio, who had run for governor twice before—he lost in 1977 and again in 1981 in an extremely close race against Kean—was known primarily as a pioneer in hazardous waste regulation.[65] His victory in 1989 was based, significantly, on his strong environmental record and the questions he raised in political advertisements about Courter's handling of hazardous wastes on his own land. Hazardous waste regulation was at the center of state politics and, in public opinion polling, consistently ranked at or near the top of citizen concerns.[66]

New Jersey's severe hazardous waste problem, spectacular disasters, administrative scandals, and state politicians ready to champion the cause of tough regulation all combined to produce a classic example of the role of entrepreneurial politics in social regulation.[67] Even before the Love Canal incident, publicity over leaking landfills pushed hazardous waste regulation onto the state's public agenda, and, in 1977, the legislature passed the pioneering Spill Compensation and Control Act, commonly known as the Spill Fund Act. This act, which taxed the petroleum and chemical industries in the state to pay for the cleanup of leaking hazardous waste sites, is important historically because Florio used it as the model for CERCLA. But the act's redistributive funding mechanism is theoretically interesting because it challenges a key assumption we have made about the ability of states to pursue such policies.

New Jersey's encounters with hazardous waste "events" only increased in the 1980s. The Chemical Control explosion occurred in Elizabeth in April 1980, and its aftermath created a highly publicized scandal. Cleanup of the site cost $33 million ($30 million from the state's Spill Fund and the balance from Superfund). Political pressures for reform mounted when investigative journalists, particularly those writing for the *Newark Star-Ledger* (the state's leading newspaper), disclosed that organized crime had connections to Chemical Control and that there was ineptitude on the part of DEP.[68]

The drama of the events in Elizabeth and the revelations of administrative bumbling and criminal involvement promoted hazardous waste issues as an entrepreneurial phenomenon. The state legislature responded with two pathbreaking laws: the Solid Waste Management Act of 1980 (facilitating the state's compliance with RCRA) and the Major Hazardous Waste Facilities Siting Act of 1981 (which we discuss in chapter 6). In 1983, events cooperated again, as the discovery of dioxin contamination at the Diamond Shamrock site in Newark caused enough public turmoil to encourage the legislature to pass the powerful Environmental Cleanup Responsibility Act (ECRA) and the controversial Worker and Community Right to Know Act (which we discuss in chapter 7). During our interviews in 1983, just after the passage of these acts, Assemblyman Ray Lesniak, who championed both bills in the New Jersey legislature, confided, "If it weren't for that dioxin site, we might have lost one of those bills!" The accidental release of an eighteen-mile-long toxic cloud from the American Cyanamid plant in Linden in October 1984 led to the passage of the Toxic Catastrophe Prevention Act in January 1986.[69]

DEP personnel associated with the agency's mismanagement at the Chemical Control site were replaced, and a large-scale reorganization occurred, the primary goal of which was to focus greater agency resources on hazardous waste regulation and disposal.[70] This reorganization and several subsequent ones greatly improved the agency's reputation in the eyes of the legislators most closely associated with the hazardous waste issue; but the changes did not necessarily translate into improved public relations for DEP. Our informants in the legislature, in environmental groups, and in DEP itself contended that widespread public dissatisfaction with the agency continued to exist. In a manner reminiscent of the disparity between EPA and citizen rankings of the hazardous waste issue at the national level, our DEP sources characterized public attitudes in a managerial fashion, suggesting that public perceptions of ineffectiveness are less a function of bureaucratic ineptitude than a matter of public sensitivity to hazardous waste issues in their communities and the sheer size of the regulatory task confronting DEP.

This very problem of magnitude draws us to regard the levels and sources of funding tapped to deal with the issue. In 1983, virtually all of our informants expressed pessimism over the ability of the state to follow through on the funding commitments made in the Spill Fund Act. Owing to the act's funding mechanism, little money had been drawn from the fund to clean up waste sites. The Spill Fund was supported by a tax on the petrochemical industry that was designed to raise approximately $5 million per year. If cleanup expenses exhausted the fund, the tax would double to ensure sufficient future funding. Arguing that it was already overburdened by the state, the industry lobbied hard against using the fund liberally, to the point of challenging the law in court. Because New Jersey's economy was still sluggish at this point, the petrochemical concerns could argue persuasively that they should not be forced to assume additional tax burdens. Such arguments were reinforced by the closing of the Hess refineries and the diversion by Exxon of new plant investment to Louisiana—a state, an industry lobbyist noted, that was not known for either stringent regulations or high taxes.

Our argument about redistributive tax burdens is nicely illustrated by industry's behavior at this point. Notwithstanding the growing public alarm about hazardous waste at this time, the entrepreneurial politics responsible for the tough legislation mentioned above was not sustained in appropriation schemes. In fact, the threat of industrial exit prompted state legislators to authorize a $100-million general obligation bond issue in 1980 to supplement the Spill Fund for cleanup of the state's hazardous waste sites. These bonds, to be retired out of general state revenues, effectively socialized the costs of cleanup by shifting state funding mechanisms from the industry responsible to the taxpayers in general.

As we mentioned, the politicoeconomic complexion of New Jersey changed by mid-decade. And with it changed the state's willingness to protect its petrochemical industry. Diversification of the state's economy removed its heavy dependence on the older industries. The state's image as an ecological disaster area, bolstered by the dramatic stories described above, turned effective regulation into an economic development issue because, the state argued, it would be difficult to attract cleaner industries and their middle-class employees unless the state got tough on toxics. Thus, funding increases for hazardous waste regulation began to come from increased taxes on polluting industries (in a redistributive fashion) as well as from a growing tax base (in an allocational fashion).

New Jersey routinely appropriated major resources for cleanup later in the decade. This money was raised both from general tax sources and from increased taxes on the petrochemical industry. Reflecting the magnitude of the state's economic turnaround, the legislature was able to draw $100 million for hazardous waste cleanups out of a $700-million budget surplus in 1985. One year later, as part of another reorganization of DEP, the legislature passed a funding package which, including $500 million from federal Superfund payments and enforcement actions against responsible parties, provided about $1.6 billion over five years for the cleanup of 229 of the state's estimated 600 hazardous waste sites (including the 97 then on EPA's NPL). Of this money, $335 million came from new taxes on the oil and chemical industries and a general corporate franchise tax. Three hundred million dollars came from a bond issue passed in a popular referendum in October 1986—by a two-to-one majority of the state's voters.[71]

DEP's implementation of the state's hazardous waste regulations was clearly affected by the infusion of moneys and strong legislative and gubernatorial support. Though still a controversial agency in the public's eyes, DEP was prepared for Superfund expenditures when they became available for the first time in 1983. The state received 40 percent of the Superfund money spent nationally during that scandal-plagued first year of the program, precisely because DEP—alone among state environmental agencies—was ready to spend the money and knew that the EPA was interested early on in funding only site-planning activities rather than actual cleanup. New Jersey's enforcement record has been much stronger than that of its neighboring states. In a report filed by the Northeast Hazardous Waste Project in 1986, New Jersey was found to prosecute more frequently and to target larger firms handling bulk amounts of hazardous waste for disposal, rather than the small generators that the other states often went after.[72]

The strength of legislative support for effective regulation was also evident in the passage and implementation of ECRA, the law requiring that industrial property be certified clean of hazardous substances before title can be transferred. Interestingly, in 1983, ECRA passed without much publicity or opposition because industrial lobbyists were busy contesting the state's right-to-know act, also under deliberation at that time. But ECRA has turned out to be a strongly redistributive piece of legislation. Sources in the legislature indicated that industry has paid out enormous sums of cleanup money without any prodding from state enforcement officials. ECRA and its implementation have been hailed by the Council of State Governments as a model of innovativeness and effectiveness.[73]

By decade's end, politicoeconomic forces in the state had produced a large, durable majority supporting tough legislation and the funds for implementation. The new, bipartisan acceptability of redistributive funding in New Jersey was illustrated by Governor Kean's outspoken support for new taxes on polluters—surprising for a prominent Republican. The New Jersey experience is pivotal because it highlights the constellation of forces required to overcome the structural limitations on state action: a severe, well-publicized problem, the specter of public scandal, and a strong, diversified economy.

New Jersey was, in essence, a best case scenario for states' abilities to deal with the hazardous waste problem in a time of shrinking federal commitment. But the hazardous waste program continued to draw substantial funds, in an allocational fashion, from the New Jersey general revenue fund.[74] Where similar conditions exist, states are likely to have the capacity to address the problems posed by social regulation. Where they are absent, as in our other two cases studies (and, we suspect, in most states), the outlook is much less optimistic.[75] Further, our coverage of New Jersey's performance leaves in question the federal government's response to the state's tough regulatory program, not to mention public reaction to key aspects of that program. We address both questions later on.

Ohio

Although it too faced severe hazardous waste problems, Ohio had struggles not encountered in New Jersey and Florida, as the state grappled with the severe economic decline endemic to the Great Lakes states during the 1980s. The state's economy remained dominated by a small number of large, aging industrial concerns hit hard by the recession at the beginning of the decade. At the depths of the recession in October 1982, as we were beginning our investigation, statewide unemployment was at 13.8 percent, and for cities in the depressed northeastern part of the state, unemployment exceeded 20 percent.

In effect, Ohio was caught in a difficult economic transition just when the recession descended. First, as employment in the state was gradually shifting from its traditional heavy manufacturing base to nonmanufacturing, service sector jobs, the nature and size of the labor force changed radically. The entrance of women and young people born in the 1950s and 1960s into the employment pool increased the labor force by 20 percent in the ten years preceding 1982. But as Ohio's recession stretched from the 1970s into the 1980s, traditional job sources in manufacturing declined, and the state's fledgling service sector was too weak to take up the slack. Second, the low productivity of Ohio's aging manufacturing plants combined with relatively high union wages to place Ohio at a competitive disadvantage vis-à-vis the new, high-technology economies appearing in other regions.[76] Like its Great Lakes neighbors, Ohio was losing capital investment as well as population to the so-called New South. This, of course, made the threat of industry exit frighteningly real to policymakers. The combination of these two factors produced acutely negative economic results relative to the rest of the nation (and even within the Great Lakes region), and this bad economic news unsettled the balance of political power within the state.

Ohio's transitional economy was especially vulnerable because the state had never had the public capital to address the infrastructure needs of new industry. Ohio has traditionally had a very low "tax effort" (the proportion of personal income paid in state and local taxes), ranking last or near the bottom of the fifty states on that measure.[77] But in spite of its conservative fiscal structure, Ohio has long been a leader in the civic reform movement. This movement, originating in the Progressive era, sought to increase the efficiency of government by applying the wisdom of trained, nonpartisan experts. Its popularity in Ohio is evidence of the long-time fondness among state policymakers for managerially inspired solutions to public problems.[78] The combination of fiscal conservatism and a reform orientation in the Progressive mold makes for a peculiar regulatory context, particularly for social regulation. As we shall see, the difficult economic times of the 1980s, coupled with the problems of hazardous waste regulation, severely challenged the managerial orientation of the state's civic reform tradition.

Despite a tradition of Republican party dominance, the pressures of the declining economy in the 1980s began to change the political orientations of the state. Indeed, as we began our interviews, Gov. Richard Celeste, a moderate Democrat, was replacing James Rhodes, a conservative Republican. The environmental movement was well organized and becoming strong in the state. As we shall note in chapter 7, it was responsible for some pathbreaking right-to-know legislation. However, there was fear in the environmental community that economic decline, shrinking fiscal resources, and a frenzy of economic development strategies might erode Ohio's frail commitment to environmental protection.

The hazardous waste problem in Ohio is the result of a long history of industrial operation, but because the state's industries have been largely concentrated on the manufacture of durable goods rather than on chemical production and petroleum refining, the problem is not as severe as in New Jersey. Also, Ohio has a less fragile hydrogeology than either Florida or New Jersey, which means that its hazardous waste problem was not always an immediate or pervasive threat to groundwater supplies. Still, the presence of toxics in Ohio during our study ranked it in the top quartile of states with hazardous waste sites slated for Superfund cleanup.[79]

An exploration of Ohio's struggle to regulate hazardous wastes allows us not only to amplify many of the issues we raised in our discussion of New Jersey, but also to address several new ones. For example, Ohio's continued economic distress throughout the period of our study makes it a useful counterpoint to the New Jersey case. We would expect Ohio to be much less successful than New Jersey at imposing on industry the redistributive burdens of regulation. To begin our assessment, we examine first the regulatory infrastructure in place as the 1980s began.

The reputation of the Ohio Environmental Protection Agency (OEPA), the agency with the central responsibility for hazardous waste regulation, was not strong during most of the years of our case study. The agency's prestige had been damaged by Republican Governor Rhodes, who kept OEPA in check during its formative years. With the succession of Governor Celeste, hopes were high among environmentalists that OEPA might improve its reputation. But Celeste, not wanting to offend industry in a time of economic vulnerability, was cautious in making his appointments. As a result, OEPA did not take a leadership role during the early years of hazardous waste policy-making. Industry and environmentalists tended to fight it out in the legislature.

Agency leadership, however, was not OEPA's biggest problem during these early years. It was the source of its operating budget that compromised the agency's advocacy. The OEPA's funding for its hazardous waste management programs was derived from a tax on waste disposal revenues.[80] The programs depended upon a thriving waste disposal industry, thus providing the agency incentive to expand waste disposal capacity within Ohio's borders, but discouraging the agency from closing down violators. Needless to say, this funding scheme compromised OEPA's credibility with the public in general and environmental and community groups in particular.

The disposal tax never raised adequate operating revenues, but its pay as you go rationale was politically acceptable in bad economic times, unlike more severely redistributive taxes that would have targeted hazardous waste generators. But in 1985, the legislature was forced to address OEPA's revenue shortfall as part of a larger funding crisis regarding hazardous waste management in Ohio. The state's economic limitations and its consequent underfunding of its waste programs attracted the federal government's attention. The EPA threatened to withdraw the state's Superfund cleanup money (some $70 million) because the state's own cleanup trust fund could not meet the 10 percent matching requirements under CERCLA.

Forced to meet this challenge, the legislature doubled disposal fees on landfill operators and added a tax on other treatment facilities. These moves would bring in the $7 million required by CERCLA. In addition, the legislature allocated $3.5 million from general revenues to fund OEPA's hazardous waste management program, thus mitigating the OEPA's dependence upon the target of its regulations for its operating income. The troubling incentive to encourage waste disposal as a source of revenue was still there but was attenuated by an allocational infusion from general revenues. Unlike New Jersey during the latter part of our study, Ohio's economy discouraged the imposition of redistributive costs on its beleaguered industrial base. The burden of hazardous waste management was thus spread across all taxpayers, rather than concentrated on the offending industries.

The modified funding scheme, coupled with the appointment in 1988 of Richard Shank, a respected scientist with a strong environmental background, to head OEPA, signaled an unfamiliar vigor in the agency. But by the end of the decade, whether OEPA leadership could gain some respect for the regulatory program in Ohio remained an open question. The lingering connection between tax revenues from disposal and the number of HWDFs active in the state made facility siting a lightning rod for the legitimacy of the state's overall hazardous waste management scheme. Our analysis of Ohio's protracted conflict over siting addresses this issue in detail (see chapter 6).

Florida

During the 1970s and early 1980s, Floridians saw themselves as immune to the problems of industrial pollution plaguing states like New Jersey and Ohio. Aging phosphate and paper pulp industries remained sound, but the new sunbelt economy of the state was based on tourism, retirement, and such clean industries as the emerging international financial center in Miami, the aerospace firms around Cape Canaveral, and a growing high-technology corridor in the southeastern section of the state. Thus, it came as quite a surprise when the original (interim) NPL contained more Superfund sites in Florida than in any other state.[81]

The severity of Florida's hazardous waste problem had multiple origins. First, its water supply is even more vulnerable than New Jersey's. Some 90 percent of the state's population draws its drinking water from aquifers located only a few feet below porous, sandy soils. The most vulnerable of these aquifers are located in the populous southeastern and southwestern sections of the state, areas that are also home to most of the state's hazardous waste generators. Second, many of the so-called clean industries, especially those related to high-technology products, turned out to generate sizable amounts of hazardous wastes. Third, Florida's hydrogeology made land-filling and deep-well injection—the "affordable technologies" at the time—inappropriate in most sections of the state. Thus, safe disposal required either more technologically sophisticated operations (like incineration, recycling, or chemical neutralization) or shipping of the wastes all the way up the peninsula and out of the state to landfills in Alabama and South Carolina. Shipping was relatively expensive and, consequently, created a powerful incentive for illegal dumping—precisely in those areas most susceptible to contamination.

Coupled with the severity of its hazardous waste problem, the unique constellation of politicoeconomic forces in the state made Florida a good candidate as a case study site. The old extractive industries mentioned above and the new chemical and petroleum interests had accumulated considerable political power to complement their economic significance in the state. Even though the phosphate mines and paper mills waned during the decade as economic forces, they remained a potent political force in opposition to regulation, especially when allied with the increasing political strength of newer industries.[82]

Agriculture was also a strong political and economic force in Florida. Farming interests faced cross-pressures when it came to hazardous waste regulation. On the one hand, the state's generally poor soils and semitropical location made farming a chemical-intensive enterprise. Strict regulation of fertilizers and pesticides would require changes in farming practices likely to be resisted strongly by agricultural interests. On the other hand, publicity about possible contamination of fruits and vegetables from polluted water threatened the marketability of Florida's produce.[83]

Of course, Florida's tourism industry was enormously influential, given its size and the state's nearly exclusive reliance on a sales tax for its operating revenues. Positive publicity from tourism also played a major role in generating the substantial immigration of retirees and younger folk upon whom the growth of Florida's real estate and construction industries depend. These industries were extremely vulnerable to revelations about groundwater contamination and threats to the state's fragile environment.

Finally, Florida's many extremely sensitive environmental treasures had fostered a strong environmental movement. Environmentalists were able to form alliances with other, more powerful political forces in the state and had been successful during the 1970s at moving environmental laws through an otherwise conservative state legislature. But progressive legislation was not necessarily accompanied by adequate funding for implementation and enforcement of these laws.

When we began interviewing in Florida, the limitations of existing state regulatory endeavors were becoming painfully evident. The state had provided few resources to support hazardous waste management. For both Superfund matching moneys and funds to clean up orphan sites not on the NPL, the legislature had created a Hazardous Waste Management Trust Fund with an initial appropriation of only $600,000. A 1-percent excise tax on hazardous waste disposal was created to supplement the fund, but, given the tiny proportion of hazardous wastes legally disposed of in the state, this raised little additional revenue. In fact, revenues were so paltry and Florida's underfunded Department of Environmental Resources (DER) was so beleaguered that Florida remained the only state in the Southeast not authorized to monitor hazardous waste facilities under RCRA. Frustration with state efforts was such that large urban areas (especially the Miami-Dade County Metropolitan Government) began to develop their own hazardous waste monitoring and enforcement capacity.

Florida's water supply became the center of attention after the occurrence of near-record droughts and widely publicized incidents of toxic contamination in the early 1980s. Influential blue-ribbon studies by panels convened by the legislature and the governor[84] inspired the legislature in 1983 to produce the landmark Water Quality Assurance Act (WQAA), an omnibus bill representing the state's attempt to deal comprehensively with the impact of Florida's phenomenal growth on its vulnerable water supplies. Of particular relevance to our study was Section VI of the law, which in essence restructured the state's approach to regulating and disposing of hazardous waste.

The passage of the WQAA in 1983 demonstrated the power—and the limits—of entrepreneurial politics. Bad publicity about Florida's pollution problems, including embarrassing stories in such national magazines as *Time* and *Sports Illustrated,* created growing public concern.[85] Several politicians saw the opportunities in such public attention and quickly moved to associate themselves with the movement for environmental action. State Representative Jon Mills had primary responsibility for drafting the legislation and managing its passage on the House floor. Mills used his stewardship of the WQAA effectively and established his leadership credentials during the session for his successful bid to become speaker of the House several years later. Not to be outdone, State Senator George Kirkpatrick took the lead in formulating the Senate's response, using his chairmanship of the Senate Natural Resources subcommittee to advocate, among other things, the novel Amnesty Days program (see chapter 6).

Florida's focus on water pollution in 1983 made hazardous waste a mom and apple pie issue. Entrepreneurial politics was much in evidence as the WQAA sailed through the legislative session without much visible or effective opposition from traditionally formidable interests with the most to lose from its passage.[86] But entrepreneurial politics takes legislative action only halfway in overcoming the intense opposition of concentrated groups. The WQAA's financing reveals how the states can become vehicles for transforming the redistributive intent of federal and state legislation into programs with only allocational impacts.

In contrast to the legislature's overwhelming support for the WQAA itself, there was deep disagreement over how to fund the law. Initial proposals called for a clearly redistributive method of funding through a tax on the chemical and petroleum industries. Several legislators argued that solving the hazardous waste problem on a long-term basis required taxing the industries most responsible for the problem in order to create new revenue sources earmarked for WQAA implementation. They feared that if funding were to come only from general revenues or onetime sources, continuous funding for the program might be eroded in subsequent legislative sessions. Although 1983 was "the year of water quality," funding from general revenues could easily be cut back in following years, when the legislature's attention focused on other hot issues. In short, the WQAA's long-term implementation would be jeopardized by its having to compete with other programs for scarce state dollars.[87]

Such arguments notwithstanding, determined lobbying by the chemical and petroleum industries helped defeat proposals to tax them in order to pay for the WQAA. In fact, industry representatives admitted to saving their best efforts for this stage of the legislative process. So sharp was the disagreement over funding that it held up passage of the entire state budget until a compromise could be reached. Just as many of the WQAA's supporters had feared, funding for the WQAA was based on a combination of one-time and general revenue sources.

The entire WQAA was an impressive-looking $117 million package; but it was creatively financed. Major funding came from a one-time speedup in the collection of state sales taxes, which yielded $100 million dedicated to the sewage treatment programs that dominated the WQAA. The programs in Section VI dealing with hazardous wastes were funded primarily from the newly created Water Quality Assurance Trust Fund (WQATF), seeded with a one-time transfer of $11 million from another environmental trust fund (the Coastal Zone Protection Fund) and from annual transfers of interest from that fund. An additional $8.1 million in interest accrued from the accelerated collection of the sales tax was added to the fund. Fifteen million dollars from the fund was to be used, first, to match federal Superfund moneys and, second, to clean up non-Superfund sites. The balance would pay for the remainder of Section VI's innovative agenda (discussed in chapter 6). Thus, the ambitious program was financed without any increase in taxes on either pollution-producing industries or the general public. Budget conferees pledged to appropriate $4.6 million annually out of general revenues to cover, among other things, the added yearly expenses incurred by DER to implement the program. However, as WQAA's supporters had warned, those pledges were never honored in future state budgets.

The pitfalls of failing to design new funding sources dedicated to the WQAA soon became apparent. By 1988, the WQATF was drawn down to about $9 million and seemed about to be depleted even further to clean up non-Superfund pesticide contamination sites, ethylene dibromide (EDB, a pesticide) sites, and to support the state's Amnesty Days campaigns. In response, the legislature revamped the financing of the WQATF by adding taxes on auto batteries, a small number of solvents, and motor oil. It was estimated that the move would bring in an extra $7 million, thereby increasing the size of the fund to $16 million. Further, the legislature tapped general revenues for an additional $2 million to renew the Amnesty Days program and to fund a $1-million grants program designed to encourage counties to establish their own permanent hazardous waste collection/transfer facilities. The redistributive impact of this new revenue source was offset by the discontinuation of the Coastal Zone Protection Fund, mentioned above, and its associated taxes, which had been imposed upon roughly the same industries before 1988. So the 1988 legislation essentially established replacement funds for the original WQAA funding scheme, without meaningfully shifting the burden of paying for the regulations away from general revenues. And one area of environmental protection was sacrificed for another, perhaps more newsworthy, area.

Broad agreement over passage of the WQAA, coupled with protracted conflict over how to pay for its provisions, led to spotty implementation of the hazardous waste components of the act. In fact, the history of Section VI's implementation illustrates nicely how social regulation, especially at the state level, tends to result in mainly symbolic solutions to the problems it seeks to address. It was not until 1985 that DER received any of the new lines it needed to implement the hazardous waste components of the WQAA. Even then our informants explained that the increase in funding for these activities was not inspired by the unmet needs of the WQAA, but rather followed from the state's implementation of Federal Hazardous and Solid Waste Amendments (1984) to RCRA.[88] Paralleling the situation in Ohio, it was the pressure of federal changes in regulatory requirements, rather than passage of ambitious state legislation, that forced the legislature to increase the resources devoted to DER. Even then, of course, the effect of this increase was allocational, as it came out of general revenues.

By 1989, DER's budget for implementing the provisions of RCRA alone totaled $3 million, supporting a staff of about three hundred employees statewide. These seemingly large increases in staff and funding were spread thinly across the state, and grumbling about water quality enforcement both within and outside DER continued. Turnover at DER had always been a problem, but during our study it became critical. Salaries in DER were much lower than in comparable environmental agencies in neighboring states. For that matter, county and city environmental officials in Florida often made more than comparable DER personnel. A DER study committee (1986) found that the turnover rate between mid-1983 and mid-1985, a period coinciding precisely with the early implementation of RCRA, CERCLA, and the WQAA, reached nearly 40 percent among agency specialists. In 1985 alone, DER's engineering staff recorded a 16 percent turnover rate. In Palm Beach County, a part of the densely populated and industrialized southeastern region of the state, the entire enforcement section of the DER district office resigned in 1986![89] At the local level, county officials we interviewed were consistently frustrated with the lack of direction from DER. They complained that, owing to staff turnover, DER spent a great deal of time and resources training new local-level employees rather than enforcing the law.

This sort of policy-making left the WQAA short of its potential in the regulation of hazardous wastes, and thus it remained largely symbolic. Like Superfund activities at the federal level, state efforts were consumed in preliminary actions rather than in completed cleanups. Throughout the decade, the Florida legislature consistently repelled attempts to open new sources of revenue, either in the form of general taxation or more targeted levies. The state continued to rely on so-called creative financing, leaving effective enforcement of the WQAA in jeopardy.

Summary of Case Studies

The case studies of Ohio and Florida as well as our quantitative analysis raise doubts about states' financial wherewithal or political will to address the hazardous waste problem. New Jersey appeared to be an exception to this conclusion in the late 1980s, given its robust tax base, the defensive posture of its petrochemical industry, the intense citizen concern about the issue, and the state's reinvigorated regulatory apparatus. Here, by way of conclusion, it is instructive to focus again on New Jersey, as an exception that proves the rule about federalism. New Jersey's advantages in hazardous waste regulation set in bold relief the problems faced by the other two states and also warn of further obstacles in the path of successful regulation, especially as we approach the crucial issue of HWDF siting in all three states.

The infusion of resources into the New Jersey DEP and its improved reputation nevertheless left the agency with a formidable challenge in the long run, according to our sources there. They based their fears on two separate concerns. First, the hazardous waste problem was simply so massive that even the extraordinary allocation of resources from the state might prove insufficient. Further, given the long-term nature of the problem, and the relation between adequate funding and economic performance, the nationwide recession at the beginning of the 1990s jeopardized budgetary commitment to the program. Second, because the publicity surrounding hazardous waste had sensitized New Jersey's citizens, DEP was forced to respond to every new crisis, regardless of its magnitude or relative importance. This made it difficult for the agency to sustain its long-term programs.[90]

New Jersey's atypical regulatory actions provoked industry into attempting what in chapter 4 we referred to as a preemption strategy: they used the federal system to defeat state policies by appealing to the national level of government, thereby directly contradicting the professed aim of the New Federalism. At first, industry sought protection at the state level from the redistributive funding scheme of the Spill Fund Act but was thwarted in its appeals to the state legislature and in the New Jersey courts. Five industrial giants (B. F. Goodrich Company, Exxon Corporation, Monsanto Company, Tenneco Chemicals, and Union Carbide) then took their case in 1985 to the U.S. Supreme Court in *Exxon v. Hunt,*[91] arguing that CERCLA preempted states from taxing industries to establish their own cleanup funds.

The U.S. Department of Justice filed an amicus curiae brief in support of the industry argument for preemption. The Supreme Court favored the preemption argument in its decision but limited the impact to funds used to finance state cleanups of Superfund sites. States could still tax industry and use cleanup funds in state non-Superfund sites. The court's decision encountered an almost immediate congressional rebuke in SARA, which explicitly removed any preemptive intent from CERCLA, thus allowing states to use their cleanup funds broadly.[92]

The meaning of this episode in the uses of federalism can be appreciated in terms of the vertical and horizontal mobility we discussed in chapter 4. Industry's recourse to the courts illustrates how the structure of federalism can be used implicitly to "whipsaw" states in order to defeat the redistributive impact of a regulatory policy, thus protecting the position of powerful interests *without the government explicitly offering such protection.* From this perspective, the New Federalism's implicit goal was to remove the redistributive burden from polluting industries by assigning greater regulatory authority to the states and exploiting their structural inability to impose such burdens. If this strategy were to fail in states with robust, diversified economies, as it did in New Jersey, then the federal government could simply use the other side of federalism—preemption—to ensure arrival at the originally intended result.

The use of preemptive powers is, in theory, anathema to the New Federalism's stated goals of allowing greater policy-making autonomy at the state and local levels. The fact that preemption was used (and supported by the U.S. Department of Justice's amici brief) is further evidence of duplicity in the Reagan administration on the subject of hazardous waste regulation. Such inconsistency in policy-making[93] offers a rare glimpse of what the communitarian language and democratic wish might describe as the federal government's protection of privileged interests against the will of the people.

Hazardous Waste Regulation in the 1980s: Some Conclusions

Our review of the evolution of federal legislation on hazardous waste and its implementation combines with our aggregate quantitative analysis of the fifty states and our three-state qualitative study to underscore a fundamental point about the new social regulation, the uncertainties surrounding it, and its redistributive burden. Once the hue and cry of entrepreneurial politics—the dramatic events and the legislative limelight—pass from scene, the regulatory agenda remains at the mercy of the less visible sorts of politics thriving in administrative agencies and in the very structure of federalism.

Because the states were generally incapable of implementing redistributive hazardous waste policies, the Reagan administration's wholesale delegation meant reorienting sub rosa the political balance structured by Congress in RCRA and particularly CERCLA, when overwhelming majorities of both houses supported the goal of making industry most responsible for bearing the burden of hazardous waste regulation. The political economy of the states did to hazardous waste regulation what the Reagan administration could not accomplish bureaucratically in the face of public scandal and congressional opposition. The regulatory delegation strategy of the New Federalism was at once more subtle and more effective than the blatant early moves by the EPA to unburden industry of its responsibilities for the hazardous waste problem. Either way, it spelled regulatory relief.

As we move to address their role explicitly in chapter 6, it is important to remember that the competing languages of regulatory legitimacy played a crucial role in this process of regulatory relief. The Reagan administration traded facilely between managerial and pluralist discourse in the face of a growing national hazardous waste problem. On the one hand, the impact of RCRA and CERCLA could be minimized by bottling them up in the hostile bureaucracy of the EPA, where officials comfortably concealed behavior aimed specifically at undermining congressional intent behind a facade of managerial calculations. For example, in the OMB, David Stockman could disguise his essentially political distaste for social regulation through a stifling managerial process of cost-benefit analysis. In the EPA, Anne Gorsuch could cite figures "demonstrating" remarkable progress in Superfund cleanups and argue for a cessation of the program in 1985, when in actuality only a handful of sites had been cleaned up, and that with very little help from the parties responsible.

On the other hand, President Reagan, campaigning under the banner of the New Federalism, could imbue the dismantling of hazardous waste regulation with the pluralist staple of decentralizing power to a level at which interests could be better balanced. The delegation of redistributive hazardous waste policies to the states could be justified as locating power closer to the people, when in fact the politicoeconomic structure of those same states would silently diffuse instead only the costs of hazardous waste regulation to the people—without their voices really being heard—through the allocational politics emblematic of the pluralist language.

These discontinuities of discourse and performance were not lost on those forced to bear the ultimate costs of hazardous waste disposal—especially the citizens in communities that were struggling with the consequences of toxic pollution. These people were exposed to the power relations long hidden in the political process by the legitimizing rhetoric of the pluralist and managerial languages. Their realization was expressed in the neglected communitarian voice raised in protest against the hazardous waste mess of the 1980s. Communitarian discourse could expose the sham of the EPA's assurances that the problem could be solved cheaply—through better science and with the cooperation of industry. It could also reveal the hollow promises of the New Federalism. Ordinary citizens, acting upon their distrust of government policy and inspired by the democratic wish, interpreted these ploys as the power of government being used against the people to protect industry. In response, the communitarian language could give voice to one simple solution: direct opposition by citizens to federal, state, and local efforts to resolve the hazardous waste problem by siting facilities in communities across the country.

In the next chapter, we trace how the introduction of the communitarian challenge exposed the discontinuities of managerial and pluralist discourse in the states discussed above, as each sought vainly to site hazardous waste disposal facilities—the acid test, as it were, of the legitimacy of government policies in the area. In the process, we want to demonstrate how the three languages surround social regulation with incommensurable demands that can be addressed only through a policy dialogue that at once acknowledges each language's differences and permits a democratic discussion transcendent of their limitations.

6 / Siting of Hazardous Waste Disposal Facilities and the Failure of Toxics Policy in Three States

Having explored the general context of federal and state hazardous waste regulation within our theoretical framework, we now address more specifically the relevance of regulatory discourse to the struggle over HWDF siting. This issue presented all the elements that we have emphasized in discussing the crisis of social regulation. First, it had concrete consequences for the residents, governments, and industries of geographically bounded communities. Second, these communities were nested within a structure of overlapping county, state, and federal jurisdictions. Third, siting contained a volatile mixture of redistributive decision making, scientific and technical uncertainty, disparate economic and political influence, and an aroused and suspicious public with serious doubts about the soundness of the decisions being made.

All these elements led those involved in siting decisions to "talk past one another" because they used different languages to characterize the issues, and each language conceives the conflicts differently. Owing to this failure to acknowledge the differences, debate became polarized, making it impossible for the participants to find any common ground or, more appropriately, any common language with which to resolve conflicts over siting HWDFs.

In this chapter, we pay close attention to the disparate ways in which the parties justified their actions, articulated their demands, and expressed their criticisms. In many cases, managerial and pluralist languages were relied upon, particularly by government officials and established interests, to express concerns perhaps better captured in less-appreciated communitarian terms. We also assess how political institutions and processes—those existing and those proposed—affected the prospects for public dialogue and the clarification of arguments about social regulation.

A Simple Typology for HWDF Siting Processes

Any study of HWDF siting should begin by acknowledging just how difficult it was for states to accomplish the task during the 1980s. A survey coinciding with our study indicates that, out of eighty-one applications for siting HWDFs in twenty-eight states since roughly 1980, only six led to operating HWDFs by 1987.[1] Why? The answer is complicated but begins with the recognition that HWDF siting is a political process—one that varies across jurisdictions. Here, we introduce a typology of state siting processes that sorts them according to their conceptual underpinnings.[2] Simply put, we argue that these processes are inspired by differing mixtures of managerial or pluralist assumptions about regulation. Yet these languages, either alone or in combination, fail to deal comprehensively with siting as a social regulatory issue. As a result, they cannot resolve the conflicts embodied in siting.

In general, state siting programs range from state initiated, proactive siting to privately initiated, reactive siting. We refer to the extremes of this range as the *comprehensive* strategy, on the proactive end, and the *case-by-case* strategy, on the reactive end. New Jersey and Ohio represent these extremes fairly accurately, while Florida's process is something of a hybrid.[3]

States following the comprehensive path, like New Jersey, start from managerial assumptions. They typically empanel a blue-ribbon siting commission made up of state environmental, industrial, and political leaders and experts. They analyze the state's hydrogeology, industrial needs, and patterns of residence in order to determine the "best" sites for HWDFs in the state. Public hearings are then held around the state to determine the suitability of these preselected sites, and finally, waste disposal firms are either asked or permitted to bid on specific sites. But the hearings are likely to be considered perfunctory because siting decisions are seen as being inherently complex and scientific and thus better left to technical experts in order to serve a larger public interest.

One problem with this approach is that designation of preferred sites occurs at the state rather than the local level. Grassroots environmental and community groups typically become skeptical of the commission's designation process precisely because of its elite representation, its heavily technical emphasis, and its removal of preliminary decision making from the affected localities. The process also tends to be time-consuming, thus offending hazardous waste generators and potential facility operators. The focus on comprehensiveness, technical information, expertise, and ultimately the belief that a list of best sites can be developed is symptomatic of assumptions deriving from managerial discourse. As such, the strategy gives short shrift to the inherently political and scientifically uncertain nature of siting decisions, thus diminishing the acceptability of the commission's decisions for all parties involved.

States pursuing a case-by-case strategy, like Ohio, generally allow for expanded review of HWDF site applications initiated by private disposal firms before a statewide siting panel. This process typically involves the affected localities only after the application by the private party has been made. The siting panel receives evidence from the site applicant and from the affected localities and then makes its decision based upon the record of the hearing. Expertise is an important and valued part of the process, but the legitimating assumption is pluralistic—let the market suggest a course of action; provide an opportunity for objections; if none is successfully raised, then the course of action is, by definition, legitimate.

This process has several advantages. First, it requires no real preparation or initial expense by the state. Second, it is flexible because the site search is market-driven. Finally, it produces quick results. But the residents of communities involved in a case-by-case siting process often feel that the disposal industry purposely targets communities that are poorly organized or otherwise unprepared to challenge the site applicant at the hearings before the siting panel. The pluralist tendency to ignore the substantive inequity of pitting politically unorganized and technically unprepared localities against a well-organized and technically creden-tialed disposal industry undermines the acceptability of the panel's judgment.[4]

Given that both of these approaches to siting are likely to provoke distrust in the target communities, are there alternative processes that might produce more acceptable results? According to Michael Elliott's research,[5] technical guarantees of safety count for less than substantive public participation in the siting process. Elliott constructed a simulation in which community representatives likely to be players in HWDF siting decisions were confronted with proposals from three mythi-cal waste management companies and asked to rank them in preference order. All three proposals involved the same site and equivalent expenditures. Two compa-nies stressed technological approaches to risk management, one emphasizing ad-vanced risk prevention technology and the other stressing advanced risk detection technology. But the third proposal stressed completely open management of the company's site through a safety board composed of community residents.[6]

Elliott found that nearly half of the simulation participants preferred the third proposal to either of those emphasizing technical safety. The implication is that legitimacy may be increased by community involvement in facility management rather than by the delegation of management to experts. If both siting and facilities management could be structured so that residents and facility operators share an interest in the safe disposal of hazardous wastes, then, Elliott argues, the "coproduction of safety"[7] might emerge. This requires open processes, wide access to information, and adaptation to community demands, all undergirded by extensive liability provisions, so that the facility operator's interest in safety merges with that of the community. This sort of substantive participation by citizens might establish the "community of interest" necessary to make the redistributive consequences of siting decisions more acceptable.[8]

The suggestion implicit in this work is oblique to both the managerial and pluralist approaches to facility siting and operation and the ways they are com-bined in siting procedures. It makes sense only from the less conventional communitarian understanding of democratic power. Joint management of HWDFs means directly involving those members of the community who are affected most by the operation of the site. This feature, of course, challenges conventional assumptions about public power and private property and suggests rethinking the relation between market and government—topics that are off-limits in managerial and pluralist discourse.

The issue of site selection and management might better be addressed as part of a broader public dialogue featuring continuous interaction among the state (playing the enforcer of liability provisions in the dialogue, for example), the facility operator (bringing diminished expectations of property control and profitability to the dialogue), and residents at the grassroots (gaining from the dialogue a heightened sense of social responsibility for safe waste disposal). Under these conditions, enough trust among parties[9] might be produced to allow for successful HWDF siting.

The Politics of Siting in New Jersey, Ohio, and Florida

Our typology stresses the shortcomings of state siting efforts in the 1980s generally, and these shortcomings are abundantly illustrated in the experience of our three case studies. New Jersey's relative success with regulating hazardous waste was matched only by its frustration at failing to site any HWDFs after more than a decade of painstaking work. Ohio, initially embarrassed by its quick fix approach to the permitting of HWDFs, emerged with a better understanding of the issue, although no better prospects for siting new facilities. And Florida began with a highly promising experimental approach to siting HWDFs but flagged in its commitment to change the politics of siting when it actually came to locating a facility, thus leaving participants feeling confused and betrayed.

New Jersey

The struggle to site HWDFs in New Jersey has been a long one.[10] Over a decade ago, the state, sensing a need for more disposal capacity, moved from a case-by-case approach to a comprehensive siting process that could override local veto of the selected sites. But the change brought only frustration, resulting in a failure to site HWDFs anywhere in the state. This failure is not surprising, given our discussion above, but what is surprising is that the state might have been more successful if it had simply learned from earlier siting efforts within its own borders. David Morell and Christopher Magorian's description of these efforts[11] provides a useful starting point for an evaluation of the process that New Jersey eventually adopted.

Between 1976 and 1978,[12] in the city of Bridgeport, the Monsanto Corporation successfully sited a hazardous waste landfill to handle chemical sludge from its

Delaware River Plant. The company sited the facility on its own property, and it was approved without any appreciable local opposition.[13] Morell and Magorian suggest that the siting proposal was successful because the company cultivated the trust of the local citizenry. First, Monsanto publicized its plans early and avoided the appearance of trying to "slip one by" the local community. Second, the company worked closely with local officials.[14] Third, Monsanto was an important employer in the area and had good standing as a part of the Bridgeport community.[15] Both the company and the city had a direct and tangible economic stake in the continued operation of the plant and in the safe disposal of the wastes it produced. Finally, and most important, Monsanto negotiated over an extended period of time the specific details of facility operation and its safety plans. As a result, plans for the facility changed substantially to reflect community concerns.

Events at Bridgeport contrast sharply with the failure in 1979 to site a regional hazardous waste disposal facility in Bordentown. The failure in Bordentown can be explained in several ways. First, Earthline (an outside company) sought to site a hazardous waste landfill that would serve the entire region, not just local disposal needs. Second, the facility would have brought few economic benefits (for example, jobs, tax revenues) to Bordentown. Third, Earthline's parent company, SCA Corporation, had a history of troubled relations with the town.[16] Fourth, Earthline adopted a confrontational approach toward the community, challenging local opposition at meetings and in the media. Fifth, local officials felt pressured to make a quick decision about the site. Amid the struggle, a study by New Jersey's Stevens Institute criticized the Earthline proposal, and the company was forced to drop its siting effort.

These contrasting cases illustrate several characteristics of successful siting: first, meaningful local involvement in the siting process and facility operation; second, the need for trust between facility operators and local residents; third, the prospect of deriving local economic benefits from the facility; and, finally, making the siting process legitimate in democratic as well as technical terms. All of these aspects of siting are suggested by our dialogic model of policy-making and its advocacy of negotiation from positions of equal power and understanding. Although these cases hardly offer definitive lessons, they are a useful counterpoint to the siting path chosen by New Jersey in the early 1980s.

After a leaking landfill in Jackson Township contaminated the community's water supply in 1979, siting HWDFs in New Jersey became increasingly difficult.[17] As the pressures to site facilities increased and the failings of the state's case-by-case process became obvious, the legislature sought to change the procedure. In 1981, Senate Bill 1300 became law and created a new siting process giving strong preemptive powers to state government and conforming to the comprehensive approach. The bill created two new bodies, the Hazardous Waste Facilities Siting

Commission (HWFSC) and the Hazardous Waste Advisory Council (HWAC) given the responsibility, along with DEP), for siting hazardous waste disposal facilities. The managerial inspiration of the siting process was tempered by some concessions to pluralist logic. DEP was assigned the managerial task of developing objective criteria for siting. The HWFSC, a pluralist body of representatives from various organized interests concerned about the siting issue, had the job of implementing those criteria.

The first stage in the new procedure called for DEP to develop the formal siting criteria to be used by the HWFSC. This was treated as an apolitical process, consistent with the governor's Hazardous Waste Advisory Commission's call for "objective selection criteria."[18] Two public hearings on the criteria attracted little attention. The results were favorable to the location of HWDFs in virgin (or so-called green fields) territory, far from industrial and urban centers (brown fields). These allegedly objective criteria later became enmeshed in a green fields–brown fields debate that led to the collapse of New Jersey's siting process at the end of the decade.[19]

The HWFSC was responsible for using the DEP criteria to identify locations most appropriate for HWDFs. Consistent with pluralist logic, the composition of the commission, whose members were appointed by the governor, was designed to balance the organized interests arrayed around siting. It consisted of three local officials, three industry representatives, and three representatives of environmental or public interest groups.[20] When a specific site was being considered, two additional representatives from that area were to be added, namely, a county official and a municipal government official.

The HWAC, composed of thirteen members appointed by the governor, had less clearly focused responsibilities but was charged with providing additional public input during the siting process. As with the siting commission, its members were state and local officials and representatives from organized interests involved with siting issues—again, a nod to pluralist standards of legitimate regulation.

Once sites were designated by the HWFSC, the affected municipality had six months to do its own siting study. Funds for this study, including the hiring of trained personnel—called rebuttal experts—to advocate the municipality's interests, were supplied by the state. Following the completion of the local siting study, a state administrative law judge was to conduct an adjudicatory hearing with the affected municipality as a party of interest. In this hearing, the judge could find the site acceptable only if it would not be a "substantial detriment to the public health, safety and welfare of the affected municipality."[21] After the administrative hearing, the HWFSC was to make a final ruling, which was subject only to judicial review.

In the end, for all its concessions to the pluralist model, the New Jersey process remained essentially managerial, giving strong preemptive powers to the siting commission and its objective criteria,[22] a point reinforced by the comments of the HWAC's chairperson. His blunt admission of hostility to public participation reveals a strong managerial bias:

> The New Jersey siting plan is the only way to get siting done. First, you have to accept that there is a risk, then look for ways to minimize that risk in case of an accident. You don't get the public involved. These environmental groups all just want the stuff shipped to South Jersey. Sooner or later you have to accept the consequences of industry.

Writing before the process was put into operation, Morell and Magorian accurately predicted the problems that would develop—problems that were revealed in our own research in New Jersey. They note that representation on the boards was limited to organized interest groups, especially industry trade associations, state-level environmental groups, and state and local government officials. Such a scheme of representation neglects the degree to which previously unorganized citizens become mobilized and organized as the siting process develops over time. These new groups reject the assumption that their interests were being represented either by existing interest groups or by government officials.

Such rejection illustrates the weaknesses of the pluralist model of representation as applied to the issues of hazardous waste management. In particular, pluralist discourse ignores the structural barriers to organization and participation facing many segments of the populace affected by regulatory decisions. As we discuss the demise of the siting process in New Jersey, it will become clear that this problem is responsible for the split between state environmental groups, which were well represented in the process, and local citizens' groups, which felt excluded. Interestingly, the state groups increasingly adopted managerial language to express their understanding of the siting process as they became more identified with it. In contrast, the local groups spoke as out-groups, in a language clearly rooted in communitarian terms.

By 1990, the HWFSC was no closer to siting a facility than it had been in 1981. Using the original criteria developed by DEP, the commission conducted hearings, identified sites, and funded local studies. By the fact of their systematic exclusion, this process mobilized citizens' groups, which were angry, first, at having no say in technical matters (which were removed from public comment by the law) and, second, at being ignored in the selection criteria, which favored organized groups for membership on the HWFSC and the HWAC. At the end of 1988, just as the commission was preparing to narrow the list of potential sites from three to one, a variety of interests began to question the adequacy of these supposedly objective criteria. Thus, as the HWFSC was, in the words of one of our informants, "a hundred yards from the finish line," the entire process began to unravel.

In effect, the elaboration of criteria that had been categorized as technical—

therefore not requiring extensive public participation—was being called into question as essentially political. Over the years, citizens began to realize that there was no technical or objective procedure for deciding whether facilities should go in brown fields or green fields, and that any set of criteria would arbitrarily favor one location over the other. The decision about appropriate criteria was, of course, a political decision about how to distribute the costs, risks, and benefits of hazardous waste regulation. In fact, our legislative sources suggested that the original bill passed because legislators from the more populous areas of the state were convinced that the criteria to be developed would keep HWDFs away from their constituencies. Groups representing green field areas, which mobilized in response to the ongoing siting process, began to see that their interests had been reflected neither in the formulation of criteria nor in the membership of the boards that applied those criteria.

Our interviews in New Jersey revealed the dynamics of division within the environmental camp over this issue. State-level groups, well established and representing an educated, elite constituency, had been involved in the process from the beginning and were formally represented on both the HWFSC and HWAC. Adopting a statewide perspective, they accepted both the pluralist assumptions behind their inclusion on the commission and board and the managerial perspective implicit in DEP's development of siting criteria. This understanding is reflected in Morell's and Magorian's interviews with representatives of state-level environmental groups. Note the pluralist underpinnings in their summary of one interview:

> Diane Graves of the New Jersey Sierra Club asserted that if any 500 individuals were to participate in a siting process, they would select leaders and sort themselves out into groups. Rather than taking the time and effort to have this "sorting out" process occur, however, she considers it easier to rely on existing organizations. . . . In turning to these groups, Graves feels that one can "safely assume" that the different siting interests will be adequately represented.[23]

When we interviewed her in 1983, Graves was at least sensitive to the charge that the Sierra Club had become separated from the grassroots because of its representation on the HWFSC, and she was concerned about the suspicions of the local groups. But more recently, she argued that the statewide perspective of many environmental groups was the strong suit of New Jersey's siting process. Using the logic of pluralism to equate the preferences of organized and formally represented groups with the public interest, she defended the state's comprehensive approach to siting:

> Because of the process, *and because environmentalists recognize the need for these new facilities,* there has been no joining of the Sierra Club or any of

the other statewide or regional groups with the local groups that have, understandably, opposed facilities (emphasis added).[24]

Other informants were more preoccupied with the rift between state and local groups and the effect it might ultimately have on the environmental movement in the state. Jim Lanard of the New Jersey Environmental Lobby raised questions about this following his inclusion in an informal Hazardous Waste Site Mitigation Advisory Group, established by DEP to enhance communications between environmentalists and the agency. We attended the first meeting of the group, and nearly its entire agenda was occupied with the "local inclusion" question. After the meeting, Lanard commented,

> Why do we have this group? Why were these people chosen [mostly statewide environmentalists with a few of the more established local groups represented]? I'm worried about the political costs of being on this Advisory Group. I feel like we are dividing insiders and outsiders, like the Ironbound Community. . . . I want "single-issue" people in the Group, so that I and DEP can get the community perspective on things. I don't care if they don't want to cooperate [with the siting process].

These doubts about the adequacy of managerially inspired criteria and pluralist representation were repeated adamantly in our interviews with local community groups. These groups rejected the notion that the criteria developed by DEP were objective and argued that the bias toward granting representation on the boards to organized interests and elected local officials systematically excluded many unorganized and, therefore, unrepresented interests.

One of the more interesting groups we visited was the Ironbound Community Corporation of Newark, referred to above and in chapter 1. The group organized the neighborhood of a heavily industrialized and poverty-stricken area of Newark in the late 1960s. At first, the Ironbound group was a day-care provider and "general social services center," according to its members. It expanded the social services aspect into legal and welfare rights advocacy, focusing on neighborhood problem solving. In Newark in the early 1980s, the "neighborhood problem" turned out to be hazardous waste.

Although the initial DEP criteria favored a green fields location, Ironbound members were skeptical, believing that in the end the process would designate sites like Newark for HWDFs. Their situation was complicated by some of Newark's elected officials, who, at least initially, saw HWDFs as an economic development priority, arguing that such facilities would bring jobs and investment to the economically depressed city. Controversy swirled around requests for a permit to allow At-Sea Incineration, Inc., to use Newark as the home port for its ships, which are designed to burn hazardous wastes at sea.

Members of the Ironbound Community fought the proposal on a variety of grounds, first, challenging on technical grounds the safety of the proposal and, second, arguing in communitarian terms that it was unfair to force a city already devastated by industrial pollution and economic depression to bear the additional risks posed by such a facility for the sake of a few jobs. In other words, they felt that economic desperation was an illegitimate rationale for the allocation of the risks of hazardous waste disposal. Further, they saw themselves as representing the interests of the disenfranchised—the poor and mostly African-American residents of the city who were unorganized and unlikely to participate in local politics or find a voice on the HWAC or HWFSC. One member claimed,

> We want to establish a citizens' siting commission, composed of the people, elected officials, and scientific types left off the HWFSC, and then we are going to use public relations and the media to get the public to choose between the official siting criteria of the HWFSC and the "citizens" criteria.

The communitarian content of this remark could not be more clear. In short, the Ironbound Community identified the structural barriers to participation that rendered inadequate the pluralist representational assumptions behind the two boards.

In similar fashion, they distrusted the state-level environmental and DEP officials for their managerial attempts to transform essentially political issues into technical ones. Again from our interviews, an Ironbound member asked,

> How can environmentalists back these things [for example, the chemical industry, the alleged Mafia connections in the hazardous waste disposal industry]? DEP, the national environmental groups, and consulting firms are just a big buffer between the chemical industry and the people. We are trying to attack the legitimacy of DEP and the siting process. They always trade on the "knowledge gap" between locals and regulators. Well, it's not that great a gap once you get involved.

The Ironbound Community was particularly incensed in 1983 about the shift in disposal technology toward incineration, and their active opposition to the At-Sea Incineration proposal symbolized their concern. They felt that the statewide environmentalists were too concerned about water pollution and not concerned enough about the air pollution that would result from mass burning of hazardous wastes in New Jersey. But their opposition went beyond scientific disagreement to engage the communitarian issue of public power and private property. They felt that resisting the siting of HWDFs would force industry to rethink production processes, but that there were limits to how far industry would go. Conventional politics and technology combined to make it impossible to solve the hazardous waste problem, according to this member:

So you either have to redesign the production process to reduce hazardous waste output or, if you can't redesign, then you simply stop production. Economically, we can't do that under present ownership.

The polarization of forces in the hazardous waste siting issue did not bode well for the siting of HWDFs in New Jersey. Like any comprehensive strategy, the state's approach was managerial in inspiration, ignoring the political content of essential aspects of its siting process. Even with its efforts at pluralist representation in the process, New Jersey's failure to consider the communitarian implications of excluding local grassroots groups made conflict at the local level inevitable. Local inclusion in the process was essentially after the fact, and then extended only to elected officials. The specter of preemption and the lack of any real movement by the Siting Commission to educate, inform, or include the people potentially affected by their siting decisions guaranteed resentment at the local level. Policymakers neglected their own experience with a successful and inclusive siting process. They opted instead for a managerial one with pluralist trimmings. Both of these legitimating languages were overwhelmed by the communitarian objections of those excluded from decision making at the local level.

Ohio

We mentioned in chapter 5 that HWDF siting is the key to understanding Ohio's overall regulatory effort during the 1980s. The state's civic reform tradition, with its managerial underpinnings, combined with severe fiscal constraints to influence Ohio's formative struggles with siting. Early on, regulators underestimated and even scoffed at the relevance of popular participation in siting decisions, while confidently relying upon the soundness of the hazardous waste disposal industry as a revenue source for Ohio's regulatory program. The state seemed determined to authorize as many HWDFs as possible in order to expand the tax base for hazardous waste management. But lessons learned by the end of the decade encouraged those involved to embrace a more inclusive approach to siting, as events forced them to question older, discredited technologies for disposal and to consider other alternatives.

When our investigation began, Ohio's policymakers were confident that they had a workable solution to the hazardous waste crisis. First, they believed that land-based disposal of hazardous waste was safe enough, given existing disposal technologies and scientific information about Ohio's hydrogeology available at the time. Second, their plan to tax HWDF operators would make their regulatory program fiscally self-sufficient at a time when the state's dismal economic climate made other new taxes or tax increases unpalatable. So they moved quickly to create a process that would allow accelerated siting of secure HWDFs. Timed to coincide with federal implementation of RCRA and CERCLA, Ohio's siting efforts were aimed at taking advantage of the state's central location in the industrial heartland. Policymakers felt that the state's ready capacity would generate revenues by attracting wastes diverted from facilities closed down by RCRA and generated by Superfund cleanups.

One thing stood in the way of this plan, an "obstacle" revealingly identified in a law review article written in 1980 by James F. McAvoy, then director of the OEPA. Considering the author's official position, the article is especially interesting because it is such a clear articulation of managerial thinking about hazardous waste management, comparable in many ways to the perspective of the chairperson of New Jersey's HWAC, quoted above. McAvoy grounded his arguments confidently in "scientific expertise," declaring the suitability of the state's hydrogeology for land disposal, but then deploring the lack of approved sites. Standing in the way of approval was too much public participation in the siting process:

> The citizen appeal provision of OEPA's enabling statute could also prove to be a major obstacle. Under this law, any person or public official acting in a representative capacity who states that he would be aggrieved or adversely affected by OEPA's issuance of a permit or approval has the right to appeal OEPA's action. . . . It is entirely possible that several years may elapse between the OEPA's initial action on an application and the exhaustion of all appeals. The prospects of long delays tend to intimidate applicants for waste disposal permits, as soaring interest rates and rapidly escalating costs can make such delays disastrous.[25]

McAvoy's article was a timely appeal, aimed at influencing the 1980 legislative changes in Ohio's HWDF siting procedures, in which existing provisions for citizen participation in siting decisions were being targeted for elimination. As part of the state legislature's effort to bring Ohio into compliance with RCRA, Senate Bill 269 created the Hazardous Waste Facilities Approval Board (HWFAB)[26] and granted it strong powers to preempt local opposition to HWDFs. The original version of this bill called for some local representation on the board, but even this provision was removed from the final version. The composition of the HWFAB guaranteed permanent representation of an administrative and scientific elite: the director of OEPA, the director of the Department of Natural Resources, chairperson of the Ohio Water Development Authority, a chemical engineer, and a geologist.

In contrast to New Jersey's comprehensive approach to siting, Ohio's process defined the case-by-case end of the spectrum—a pluralist process with managerial trimmings. In approving HWDFs, the HWFAB followed the pluralist logic described in our typology, but its membership was entirely composed of experts, thus implicitly discounting the political aspects of HWDF siting and the uncertain, contested, and value-laden decisions made therein.

The simple process was to be initiated by a private disposal firm, which would apply to the HWFAB for approval of a proposed site. Affected communities were involved only after the application had been made. In our early interviews, HWFAB officials considered this a strong point of the process, arguing that public participation earlier in the process would make "technically correct" decisions impossible to implement. In their view, the deliberative nature of the comprehensive strategy was flawed. It took too much time, thus enabling citizens' groups to find out which sites were being considered so that they could mobilize local resistance. Ohio officials actually saw the after-the-fact local participation of Ohio's case-by-case process as a way to overcome the problem of citizen participation while supplying some pluralist window dressing. An open (market-driven) process of site selection and an appellate process for local objections to HWFAB approvals were considered adequate means for legitimating Ohio's siting decisions.

But Ohio's process limited its concessions to pluralist discourse. First, the HWFAB was empowered with preemptive authority over local opposition.[27] Second, the entire siting process was designed to operate out of the glare of public attention and to render rapid decisions. For example, opportunities for public participation were narrowly defined. After a firm had applied for an HWDF site permit, a single administrative hearing was to be held within 60 to 90 days of the firm's application, and then an "adjudication hearing" between 90 and 120 days after the application. And only parties directly "aggrieved or adversely affected" by the proposed facility were to have access to the hearings. Third, judicial appeal from the HWFAB's siting decision was to be limited to parties to the original hearing and was to be removed from the localities affected and held only at the Franklin County (Columbus) state Court of Appeals—not the state's regional appellate courts. Fourth, appeals were not to stop the execution of the permit unless extraordinary circumstances could be demonstrated. Finally, the grounds for appeal were generally limited to the record developed at the hearing.[28] If the court found "substantial evidence" supporting the HWFAB's decision to site, then it was bound to uphold the board's permit.[29] By comparison, the appeals process in Ohio granted less discretion to the court than was the case in New Jersey.

In effect, the new Ohio siting process answered every criticism raised in McAvoy's article. And initially the HWFAB accomplished the objective of efficient siting with stunning success. Within one year, the HWFAB had granted permits to 336 existing hazardous waste facilities, bringing them into compliance with RCRA and assuring abundant capacity (and revenue sources) for OEPA's toxics management program.[30] One of those facilities was the CECOS/CER landfill east of Cincinnati. The story of that landfill became a symbol for Ohio's later regulation of hazardous wastes, revealing in detail the costs of Ohio's early managerial confidence in landfill technology and the inadequacy of its limited pluralist vision of regulatory legitimacy.[31]

The events surrounding the CECOS/CER landfill in Jackson Township of Clermont County and the subsequent revelations about mismanagement and leaks at the facility nicely frame the actions of citizens' groups, industry, OEPA, and the HWFAB. They also offer evidence of three factors so often present in siting controversies during the 1980s: first, the startling uncertainty of the early scientific and technical information about hazardous waste disposal; second, the problems that flow from relying upon that information while at the same time discouraging public input and criticism; and, third, the fragility of public trust in the actions of facility operators and government regulators.

The Jackson Township site began as a sanitary landfill in 1972, built by the Clermont Environmental Reclamation Company (CER), a locally based firm, at the behest of county officials concerned with the lack of local solid waste disposal facilities. Although citizens living near the dump challenged the county about its location, the absence of zoning ordinances in Jackson Township prevented any effective opposition. Such ordinances were eventually passed, but the CER landfill (hereafter CECOS/CER)[32] was effectively grandfathered in against all later zoning changes.

In 1976, CER applied to OEPA for permits to dispose of hazardous wastes. The application process occurred with virtually no press coverage or public awareness. Public notices in the local newspapers, required by law, attracted little attention, both because they were couched in terms of solid waste permit extensions without specifying the nature of the hazardous substances that would be accepted, and because, in 1976, two years before Love Canal, there was little public concern over the whole issue. In any event, CECOS/CER quietly became "the only secure landfill for hazardous wastes in Ohio licensed for general commercial use, and one of about thirty in the United States capable of handling toxic and hazardous wastes."[33]

In 1978, with hazardous wastes receiving more public attention, the local media revealed that the CECOS/CER facility was accepting hazardous wastes from cleanup sites in Kentucky and West Virginia.[34] Following ad hoc protests against the facility's activity, local citizens formed I-CARE—Independent Citizens Associated for Reclaiming the Environment—and began raising questions about groundwater contamination from the site. Both OEPA and CER claimed the citizens' group was uninformed and argued that their concerns were groundless, given the advanced technologies used at the site.[35]

In our view, the response of OEPA and CER to I-CARE's fears is crucial. Instead of drawing the group into the policy process and developing an ongoing dialogue with them, officials—both public and private—dismissed the protest entirely. By assuming that their expertise was an adequate and unambiguous guide to policymaking, they missed an opportunity to educate citizens and include them in the policy process.[36] Their response, endorsed by the managerial language they used, had two effects. First, it gave the appearance that OEPA was more concerned with the well-being of CER than with that of the community. As a result, citizens came to doubt the ability of the state to act on their behalf when monitoring the facility. Second, the citizens' group became suspicious of the "expert information" upon which OEPA and CER were basing their decisions. I-CARE hired its own experts and began to attack the technical assumptions underlying CER's and OEPA's management of the site. The group's findings raised grave doubts about the technical legitimacy of the siting and subsequent management of the landfill.

In March 1979, OEPA admitted that it did not have adequate monitoring facilities to supervise the operation of the site. Soon thereafter, Jackson Township, encouraged by I-CARE, fined CER for violating the township's new zoning ordinance. CER responded by suing the township and I-CARE for $1 million. The company also named individuals in both local government and the citizens' group in the suit. I-CARE responded by countersuing the company, claiming it was trying to prevent further investigation of the site's safety.[37]

Following this legal standoff, opportunities for cooperation declined even further as CER merged with Chemical and Environmental Conservation Systems (CECOS), a national company. Then, in early 1983, the landfill was resold to the waste management giant Browning-Ferris Industries (BFI). These rapid ownership changes transferred control of the facility from a local firm to a multinational corporation distant from the community. As noted above in our discussion of Elliott's findings and the Bridgeport-Bordentown cases in New Jersey, local control seems to be one of the keys to developing cooperation and trust between the community and the facility operator; but, with successive changes in ownership, control moved further and further away from the citizens of Jackson Township and Clermont County.

During this ownership transition at CECOS/CER, the organizational structure and strategy of I-CARE also changed. The group allied itself with a growing statewide network of local environmental activists called Voting Ohioans Initiating a Clean Environment (VOICE) and other environmental and citizen lobbyists (the Ohio Environmental Council and the Ohio Public Interest Campaign).[38] What had once been a largely local issue now assumed statewide, even national, import,[39] particularly following the site's receipt of much-publicized PCB-contaminated wastes during the early 1980s.[40]

As the political clout of the citizens' groups increased, both the state and BFI took belated steps to address their concerns. More than two years after admitting that its monitoring was inadequate, OEPA stationed a full-time inspector at the site in October 1981. Once it assumed ownership of CECOS/CER, BFI developed programs to educate local citizens and released a ten-year plan to decrease reliance on landfilling and to develop other alternatives: incineration, deep-well injection, source reduction, and recycling. But these attempts to establish trust among citizens' groups, the state, and the facility were too little, too late. They were overwhelmed by subsequent events.

In November 1984, OEPA shut down CECOS/CER after the OEPA's on-site inspector discovered BFI employees pumping rainwater from the top of a hazardous waste cell into Pleasant Run Creek, which flows into the water supply of the town of Williamsburg, less than ten miles downstream.[41] OEPA officials called BFI's actions "unbelievable," and the Ohio attorney general's office sought criminal indictments against the waste firm.[42] The story made the *New York Times*, amid much public outrage. The site reopened briefly after the incident, but in May 1985, an independent hydrogeological study discovered extensive groundwater contamination around the facility. OEPA again shut down the facility, over BFI's objection that the study was incorrect. Three months later, OEPA reopened the facility but restricted its intake to non-RCRA wastes.[43]

Events at CECOS/CER forced a reevaluation of the original assumptions made by the state and by the operators about the hydrogeology of the site. The idea that the clay soils under the facility were impermeable and would protect the groundwater was challenged by additional scientific evidence. This called into question the very basis of the original preference, expressed in McAvoy's article, for landfilling in Ohio. Additionally, the discovery that many of the storage cells at CECOS/CER were leaking betrayed the faith state officials and facility operators had placed in the efficacy of the advanced landfilling technologies used there. Revelations that these technologies were mismanaged and may have polluted municipal water supplies further eroded public faith in so-called high-tech answers to the hazardous waste problem. In short, virtually every technical assumption upon which officials relied to discount citizens' criticisms of CECOS/CER was discredited, and nearly all the citizens' concerns were realized.

Bound together in one of the earliest and most radical statewide networks against toxics, Ohio's citizens' groups were developing a track record of success at exposing poorly managed sites and blocking new ones. Through their endeavors, the state's once-favored hazardous waste disposal industry declined rapidly, thus forcing, from the ground up, the realization among all parties of the need to reconsider the whole issue. Following the events at CECOS/CER, the siting process and the HWFAB became focal points of debate among environmentalists and citizens' groups on the one hand and industry and OEPA on the other. Most citizens' groups we interviewed saw Senate Bill 269 as strictly industry-inspired, and indeed, industry representatives agreed that they had worked closely with the legislature to fashion the law. As a result, citizens' groups shifted their energies from the local level and the administrative arena to the state capitol and became key actors in the legislature's reassessment of HWDF siting over the next few years, initially injecting a strongly communitarian element into the struggle over HWDF siting.

But success at the grassroots does not necessarily translate into success in the state legislature. As the debate over reforming the siting process developed, the citizens' groups' communitarian demands for democratic control over industrial risks were translated into weaker pluralist demands for token representation (for example, board membership) in the siting process. While this was a major concession, given the long-standing hostility of industry and government toward public involvement in the siting process, the muting of these demands in subsequent reforms and the state's later failure to engage the difficult issues raised by the communitarian challenge contributed to Ohio's ultimate inability to bring the siting conflict to closure by the end of the decade.

The best chance for the communication of local opposition came during the legislative session of 1983. Legislation was introduced which would have (1) added local citizens to the HWFAB;[44] (2) provided the possibility of citizen suits against HWDF operators; (3) encouraged recycling of hazardous wastes and discouraged landfilling; and (4) enhanced OEPA's enforcement authority. However, after a series of complicated maneuvers, the legislation was defeated by a coalition of industry lobbyists and legislators from the economically hard-pressed areas of northeastern Ohio. The defeat drew into sharp contrast the points of contention separating industry and their grassroots opponents.

Our interviews revealed not only disagreement over the technical and political issues regarding hazardous waste regulation, but also an articulation of that disagreement in competing languages of regulatory legitimacy. An industry representative, echoing the frustration born out of a managerial understanding of the controversy, saw siting as a technical issue that had been, for self-serving and illegitimate reasons, improperly defined as a political and legal issue:

> The problems with landfilling are legal, not environmental. It simply takes too long to get anything sited because of the laws governing siting. Mostly, hazardous waste sites become an economic concern for local property owners—they are afraid their land values will decline—not really a safety issue. With the clay base in some areas of Ohio, we don't even need plastic liners under these sites to prevent leaking.

Meanwhile, a representative from the Ohio Public Interest Campaign rejected the technical assumptions of landfilling proponents, and his argument has a strongly communitarian flavor. He saw the siting issue as a political one that had to include provisions for the expression of local demands:

> Local people should have a voice when their communities are being considered for hazardous waste facilities. The communities have no power to veto the decisions of the five people on this appointed Board. Under any circumstances, siting hazardous waste landfills is unsafe. The legislature should ban landfills in Ohio. Other states have done so.

In spite of the profound differences expressed here, rapprochement occurred in the wake of the CECOS/CER debacle, during the legislative session of 1984.

Problems with the federal administration of RCRA and CERCLA, changes in the administration of OEPA, continued problems with the HWFAB, and the increasing credibility of environmentalists and local groups forced traditional adversaries to reconsider their positions. Elites among several competing interests formed a Trialogue Group involving the Ohio Manufacturers' Association, the statewide Chamber of Commerce, the Sierra Club, VOICE, the Ohio Environmental Coalition, and OEPA.

All sides found common ground in their displeasure with the HWFAB, and their agreement led to important new hazardous waste legislation during the 1984 legislative session. Essentially, the original concern of industry with the efficiency of the siting process and the original concern of environmentalists with their lack of representation on the HWFAB produced a compromise in which the siting process was consolidated and streamlined—for example, permit renewals would not go before the board unless modifications were proposed—and OEPA was authorized to appoint local representatives to a renamed Hazardous Waste Facilities Board (HWFB)—a perfectly symmetrical managerial and pluralist exchange: efficiency traded for representation. But the legislation failed to empower local communities in the process—a key communitarian demand. It retained the strong provisions for state preemption of local land-use powers regarding HWDF sites.[45]

Although this legislation represented concessions by all sides,[46] the contending parties still had fundamental disagreements flowing from their conflicting perspectives on the very nature of regulation. Efforts at dialogue, then, served to narrow the bounds of disagreement, but because they neglected important communitarian concerns, they could not produce lasting agreement on siting. The managerial language continued to shape industry's expectations that the siting process should be efficient, and pluralist discourse continued to inform the environmentalists' concern for wider representation on the HWFB.

At first, the new and improved HWFB was given a chance by contending parties to produce better results. But the board ended up disappointing everyone in the way it proceeded with siting following the reforms. The criticisms were universal. Our last interviews with environmentalists indicated that the HWFB constantly changed the requirements for information that communities were to submit to challenge siting proposals, making it very difficult for the participants to prepare for hearings before the board. In effect, the HWFB turned the siting process into "a moving target" for participants.[47] Environmentalists and community activists also complained about the poor quality of the permitting work done by the HWFB.[48] In more general terms, environmentalists were dissatisfied with the board's unwillingness to consider alternatives to land-based disposal. If anything, industry groups were even more upset about the HWFB's performance, since no new sites have been permitted since the changes. One of our sources complained that the board simply "won't do anything," that it had consistently failed to act quickly enough despite the availability of the information necessary to make siting decisions.

By the end of our study, the mutual dissatisfaction with the HWFB once again inspired old adversaries to suggest reform: this time eliminating the board and turning siting decisions over to OEPA. Such an agreement represented a repudiation of the original basis upon which Ohio's entire siting strategy had been established. It also indicated the renewed stature of OEPA, after years of controversy, as an effective force for regulating the environment.

Although their agreement on reform produced no legislative results, environmentalists and industry representatives could be credited with forging some agreement over the future of hazardous waste disposal in the state. For example, industry had embraced the environmentalists' advocacy of the need to explore alternatives to land-based disposal. One industry lobbyist told us that new disposal sites in Ohio "just aren't going to happen. It comes to the point where it's just too damned expensive!" He concluded that waste minimization and recycling were the most viable options for hazardous waste generators.

To summarize the disputes about the siting process in Ohio, we return to the original idea upon which the HWFB's siting strategy was based. While drawing on the pluralist discourse, which trades upon the importance of adversary hearings in the establishment of legitimate policy outcomes, the Ohio process also worked to minimize effective challenges to its siting decisions. The process took advantage of the structural inequities, ignored by the pluralist language, facing poorly organized local opponents in adversary proceedings. Unaddressed was the communitarian critique of due process, which views representation of interests by elites and elaborate procedural guarantees as poor substitutes for the direct voice of the people in decisions about public risks in the marketplace.

The early permitting of numerous facilities ultimately undermined the legitimacy of hazardous waste regulation in Ohio. First, Ohio became something of a dumping ground for wastes from other states, because the HWFAB rapidly expanded the state's disposal capacity after the implementation of RCRA. Second, the continued operation of these sites, and specifically the CECOS/CER site, dramatized the degree to which uncertain scientific information and state-of-the-art technology provide a problematic basis for sound policy-making.[49] Third, citizen and environmental groups, stung by their early omission from siting and management decisions, mobilized statewide around the siting issue and lobbied, albeit ineffectively, for changes in the policy process, as evidenced by their 1983 legislative proposals.

Inevitably, the failures of the siting process and facility operations led to the formation of precisely the sort of strong local resistance to siting HWDFs that policymakers had hoped to avoid in the first place. A shared dissatisfaction with state efforts at hazardous waste management in general and at siting in particular led to halting initiatives to establish a public dialogue over HWDF siting. Unfortunately, the elite nature of this dialogue did not adequately represent the local challenge to toxics regulation. HWDF siting in Ohio ground to a halt at the decade's end still failing to acknowledge the relevance of that local voice.

Florida

Following Florida's belated recognition that it had a severe hazardous waste problem, it undertook an innovative approach to managing the problem. Central to its management scheme was finding a way out of the impasse in HWDF siting that had occurred in Florida as well as nearly every other state. But Florida's lack of secure facilities in the early 1980s was seen by policymakers as all the more critical because of the overwhelming incentives for industry in the state to dump hazardous waste illegally. When Florida found itself ranked first on the EPA's interim NPL in 1983, it had no established capacity to treat or dispose of hazardous waste within its borders. For proper disposal, hazardous waste generators had to pay for direct shipment to sites in Alabama and South Carolina, usually from the southernmost part of the peninsula. Further, the bulk of Florida's hazardous waste was generated in small quantities by small businesses that were typically running on tight profit margins. Thus, the costs of safe disposal, coupled with the remote likelihood of being caught by the state's moribund enforcement apparatus, made illegal dumping all too attractive to the average hazardous waste generator.

In spite of these widely recognized conditions, the siting process in 1983 operated under procedures derived from amendments made in 1980 to the state's Resource Recovery and Management Act (1974). While conforming roughly to the case-by-case approach, the process seemed designed to discourage private operators from proposing facilities. It required disposal firms to designate appropriate areas for facilities and propose them to the state. But in contrast to the siting process in Ohio, local governments could veto any proposed disposal facilities. As a result, between 1980 and 1983, no HWDF had even been proposed, let alone actually sited or built. Florida's failure to site disposal facilities made states like Alabama and South Carolina, which regulated the only legitimate disposal options for Florida, increasingly unwilling to continue accepting its hazardous wastes.

Fearing that the state, in the absence of adequate in-state facilities, would face a crisis in toxics disposal, the legislature in 1983 made HWDF siting the centerpiece of its pathbreaking WQAA. Notwithstanding the limitations in its funding and implementation (see chapter 5), the siting provisions of the WQAA's Section VI were highly innovative. Section VI's architects had developed a consciously considered set of assumptions about the causes of the state's failure to site HWDFs. Repeatedly, this strategy was cited in task force reports as well as in our interviews with the legislators and aides who drafted the WQAA and guided it through the legislature. All agreed that the legislation must address the dilemma of siting HWDFs when local opposition made siting a virtual impossibility.

Section VI was the legislature's solution, and its provisions were derived directly from five assumptions. First, citizens fear the siting of HWDFs without appreciating that hazardous waste will be disposed of illegally and at comparatively greater risk to their communities. Second, citizens do not believe that state government can handle safely the regulation of hazardous wastes in general or of HWDFs in particular. Third, and as a result of the first two assumptions, citizens resist the siting of HWDFs in their communities. Fourth, if citizens could be shown the extent of the local hazardous waste problem and if, simultaneously, state government could demonstrate its capacity to deal with such wastes safely, then there would be less resistance to facility siting. Fifth, encouraged by accurate information about economic demand for HWDFs and by a more educated (and therefore more receptive) populace, facility operators would attempt to site facilities, and these attempts would be more likely to succeed.

The logic behind the legislation employs an unacknowledged admixture of managerial, pluralist, and, to a limited degree, communitarian perspectives. Consistent with communitarian assumptions is the law's recognition that citizens are important to the siting process and must be educated and mobilized to support the process for it to work successfully. The pluralist influence is revealed in the assumption that citizen concerns can be accurately reflected through the representative mechanisms of government and the operation of a market economy. Yet, the legislation backed away from the full implications of the democratic wish by adopting the managerial approach to public education. It assumed that an informed citizenry would concede that experts possess the knowledge necessary to deal successfully with the problem of hazardous wastes. In other words, the law assumed that citizens, once informed, would defer to expert policymakers and simply bow out of the siting process. As we shall see, implementation of the legislation foundered on the contradictions inherent in these assumptions.

In order to implement the siting scheme of Section VI, the WQAA contained several crucial provisions. First, to increase confidence in the state's ability to regulate hazardous wastes, DER's funding and staffing for dealing with hazardous waste regulation was to be dramatically increased. Specifically, a funding increase for RCRA compliance was authorized, as were increases for the monitoring capacities of this much-maligned state agency.[50] Second, the WQAA mandated an elaborate statewide hazardous waste survey and needs assessment program aimed at determining where hazardous wastes were being generated and specifying the state's need for transfer and disposal facilities. Third, and related to this needs assessment program, the act required a small-quantity waste generator notification system to bring Florida's many small generators into the regulatory net. Fourth, each county was required to identify within its land-use plans a site suitable for a hazardous waste transfer facility. Because only a few such facilities would actually be built, the WQAA's authors believed that this would be a relatively uncontroversial way to begin the siting process. Fifth, the WQAA created Amnesty Days, an ambitious citizen education program designed to raise public awareness about the severity of the hazardous waste problem and to mobilize support for state efforts through a demonstration of DER's competence.

The law's authors thought that public awareness could be raised to a point at which citizens and industry alike would appreciate the hazardous waste problem as a shared concern. If this shared awareness developed, broad support could be mobilized for the siting of an HWDF and several transfer sites within the state. Given other states' siting philosophies at the time, such an approach was both innovative and risky. Without the campaign to educate Florida's citizens, an uninformed, quiescent citizenry might simply ignore the siting process until after the HWDF was a reality. An educational effort might awaken slumbering resistance and inspire a newly aware public to oppose siting the HWDF anywhere in the state. The bill was, thus, a gamble of knowledge versus NIMBY, and effective citizen education was seen as the key.

The primary educational vehicle was the Amnesty Days program. Under this state-funded scheme, a privately operated mobile storage and transfer caravan, preceded by well-orchestrated media fanfare, traveled to different regions of the state in 1986 and 1987.[51] In each region the caravan collected small quantities of hazardous wastes from citizens, small businesses, schools, and local governments on a one-time only, free, and no-questions-asked basis. The wastes were analyzed on-site and then shipped out of state for proper disposal. Participants in the program were given surveys to fill out (for analysis by DER) and were provided extensive information on proper hazardous waste disposal.

As a program designed to raise public awareness, rather than to solve the hazardous waste problem, Amnesty Days must be rated a success. The rate of participation exceeded expectations in its initial phases and actually increased over time, even though the mobile collection facility moved from more to less densely populated regions as Amnesty Days progressed. By June 1987, the program had visited all of Florida's sixty-seven counties and had gathered nearly 1.7 million pounds of hazardous wastes from nearly twelve thousand participants, roughly 80 percent of whom were private citizens disposing of dangerous household products (mostly old oil-based paints and pesticides).[52] Local press coverage of the various Amnesty Days visits was typically extensive and thorough in both rural and urban areas. The statewide Amnesty Days toll-free hotline averaged more than twenty-five calls a day during the operation of the program.

According to DER officials, media coverage of Amnesty Days led to growing awareness among citizens and local governments of the possibilities for recycling and exchanging hazardous wastes. This perception was shared by the program's designers. A long-time legislative staffer who worked on the drafting and passage of the WQAA concluded in 1988 that the greatest success in Florida's hazardous waste program was the successful communication of information about hazardous wastes. Judging from constituent pressures on legislators, he thought that Amnesty Days, the needs assessment program, and the small-quantity generator inventory had all served to raise public awareness of the problem of illegal dumping and of the need for safer ways to manage and dispose of hazardous wastes.

Indeed, Florida's approach to public education attracted much attention from other states. Policymakers from both Ohio and New Jersey were well aware of Florida's strategy when we interviewed them, and they remained interested in the results over time, primarily in terms of whether or not Florida's gamble would pay off in publicly acceptable siting decisions. On that crucial point, the program was much less successful because its designers wrongly assumed that public awareness would lead to public deference to the siting decisions made by elite policymakers. The designers of the WQAA implicitly shifted to a managerial frame of reference as they moved to the siting process itself, embracing what we referred to in chapter 3 as an objectivist model of scientific and technical information. They assumed that if citizens were given enough information, they would see the issue exactly the same way as the policymakers did. The resulting consensus would then legitimate adoption of objectively correct solutions to the problem and justify placing greater authority in the hands of expert policymakers.

From the perspective of the dialogic model, Florida lawmakers ignored the degree to which scientific and technical information is only one part of the HWDF siting issue. Access to such information does not lead to general agreement about policy solutions. On the contrary, the more citizens know, the less they are willing to accept decisions by experts. They come to understand the uncertainty of scientific and technical knowledge about HWDF operation, and they grasp the irreducible political components of the policy decisions made about the siting and management of these facilities. Awareness and education create a greater desire among an informed citizenry to participate in the policy process and to have their interests recognized. Public support for any policy can emerge, then, only if citizen awareness is matched by citizen access to an ongoing public dialogue out of which public policy emerges. In other words, the creation of an informed and aware citizenry is a precondition for meaningful public dialogue; it does not guarantee—indeed, it may even preclude—support for policies already decided upon by elites.

Given the history of ineffectiveness that Florida brought to hazardous waste regulation during the decade, public education and awareness were certainly worthwhile goals; however, they were only intermediate goals. Attention should also have focused upon the creation of institutional mechanisms for fostering the public dialogue that increased public awareness makes possible. And this the Florida lawmakers decidedly failed to do. In fact, as aroused and informed groups of citizens demanded greater roles and justified their claims through both pluralist and communitarian discourse, they challenged not only the knowledge base but also the political legitimacy of the state's policymakers. Unprepared for this, Florida's politicians retreated further and further into managerial rhetoric to defend their policies. The overall result was elite misuse of scientific and technical information to justify what were essentially political decisions. In response, informed citizens' groups recognized and challenged such misuse, and this led to the further delegitimation of the entire process.

To see how this happened, we need to consider the new siting process built into the WQAA and its connection to the Amnesty Days program. We reiterate here that siting an HWDF is an inevitably redistributive process: concentrated costs are imposed on one community to reduce the diffuse costs (from the illegal dumping of toxics) that would fall on the entire polity. As such, it places intense pressure on any political system, and there are appreciable incentives to avoid such decisions.[53] As was the case with increased funding for DER, the actual pressure to site at least one HWDF came less from the WQAA than from the federal government. The EPA made the facility a condition for continuing the flow of Superfund monies into Florida. This condition led to revisions in the WQAA during the 1987 legislative session, just as Amnesty Days was winding down in the rural counties of northern Florida. The legislature set May 1, 1988, as the deadline for recommending a statewide site, and it allocated $600,000 out of the Water Quality Assurance Trust Fund to do the job.

The revised WQAA directed DER to survey 10,200 parcels of state-owned land for candidate sites for the state's $50-million HWDF. Among the factors to be addressed in the selection of sites were the depth of the water table, air quality, areas designated as environmentally critical, the proximity of population, transportation access, the size of the site, local land-use plans, and economic impact on the community. In June 1987, DER paid $525,000 to the environmental consulting firm Roy F. Westin, Inc., to evaluate the parcels and recommend five candidate sites by the end of 1987. Although the legislation required no formal public participation, a series of public meetings was held during the early stages of the process, but these were not well publicized or given much media attention. Significantly, none of the meetings was held near enough to the location of the final candidate sites to attract much attention in the affected communities. Further, Westin, Inc., repeatedly failed to issue monthly "public awareness materials" and news releases, as stipulated in the consultant's contract, thus making it impossible for residents to fathom the relative priority of the sites near their communities.

On January 4, 1988, DER announced the top five sites. At the top of the list was a site in north-central Union County, the poorest and least populated county in the state. Of the four other sites, three were within a few miles of the Union County site. The only site outside of this area was DER's second-ranked location in DeSoto

County in southwest Florida. The first general awareness that the affected areas might be considered for the final list surfaced less than a month before, when Westin, Inc., issued a preliminary ranking of twenty-five sites, and, at that time, the north-central Florida sites were not ranked anywhere near the top five. Thus, there was little time and seemingly less necessity for area residents to become involved in the site selection process before DER's January announcement.

Spontaneous grassroots opposition to the selection process emerged almost immediately after the DER made the final five sites public. United Citizens against Pollution (UCAP), a group based in Union and Bradford counties, represented citizens near three of the top-rated sites. After hiring their own environmental consultants and lawyers, UCAP identified five flaws that, in their view, raised serious questions about the siting process. First, the four north-central sites had been ranked relatively low in the earlier list (the top-ranked site had been number eleven initially), which lead to the belief that political, rather than technical, criteria guided the final selection process. Second, the mayor of Raiford, the community nearest the top-ranked site, had privately encouraged DER to choose that site in November 1987, arguing that it would bring jobs into the community.[54] Third, affidavits from an ongoing Georgia lawsuit revealed evidence of political influence operating in that state's designation of a poor, rural county for its statewide HWDF (the charges were later proven true, forcing Georgia to start its site selection process over from scratch). Fourth, irregularities in the consultant's report were ignored by DER and, later, by the Environmental Regulation Commission, as discussed below. Fifth, virtually no well-publicized notice or hearing occurred until after the five sites were effectively nominated.[55]

Despite these objections and residents' request for a delay, the list of sites (now pared down to four after one of the Union County sites was dropped from the list by the consultants for technical reasons) was presented to the state's Environmental Regulation Commission (ERC) on March 17, 1987, in a public hearing to determine the state's nominee for the HWDF. The hearing took place in Jacksonville, about an hour's drive from Union and Bradford counties. Only Union County, of the four counties affected by the siting decision, failed to send an official delegation to the hearing.[56]

The ERC, a panel of seven citizens appointed by the governor and, at the time, composed mostly of business leaders and established conservationists, chose the Union County site as expected. They acknowledged but ignored irregularities in the consultant's report[57] and forwarded it to the governor and the cabinet to allow them to proceed with siting the facility. In justifying his vote for the Union County site, ERC member A. L. "Jack" Buford, a real estate developer from Tallahassee, said, "I'm comfortable that most of the technical information is accurate and reliable."[58] Other members' comments also made it clear that they saw the meeting exclusively as a matter of technical decision making. Just the day before the meeting, ERC member John Shepard, a citrus industry executive from Tampa, put it plainly: "If it comes to where two scientific experts disagree, I'll have to sit like a judge. But hopefully, there will be a clear determination."[59] After the hearing, Dale Twachtmann, DER secretary, gave the consultant's presentation to the ERC a lukewarm endorsement: "Although I'm not fully satisfied, they have done an acceptable job, I don't think it [the site selection] was fatally flawed."[60]

Given the natural resistance of any community to being singled out for such a facility and, in addition, their substantial questions about the technical quality of the site selection process, it is not surprising that UCAP challenged the site selection process both administratively and legally. Existing administrative procedures allowed for such a challenge. What is surprising, however, is the response of state officials to that challenge. Instead of recognizing that the UCAP challenge was to be expected or that it raised legitimate questions that deserved to be addressed on appeal, the state legislature intervened after the fact to halt the appeal. This move further delegitimated the state's approach to siting the HWDF and revealed a fundamental disjuncture in the state's understanding of the siting issue.

As the state came under increasing pressures from the EPA to submit its twenty-year plan for dealing with hazardous wastes (including plans for an HWDF), legislative leaders became impatient with delays in the siting process. Rather than wait for UCAP's appeal to run its course, the legislature named Union County as the site of the state's HWDF in June 1989. This legislative action left UCAP with no recourse to the administrative hearing, for which it had been preparing for over a year. The bill did provide for a special appeals process by creating a streamlined "one-stop" hearing procedure to handle site selection, permitting, and construction of the HWDF all at once. Instead of using the dispute over the Union County site as a means of recognizing citizens' concerns, the legislature by-passed existing provisions, preventing even the feeble approximation of dialogue that might have emerged from the give and take of the conventional administrative appeals process.

Legislative leaders, DER Secretary Twachtmann, and the governor, who supported the alteration of the process and the naming of Union County as the HWDF site, expressed their opinions in managerial terms. They all felt that the purpose of the process was to uncover the information required to make the one correct decision that would best serve the public interest. Thus, the siting decision was a purely technical issue, and, once the experts had identified the best site, attempts to appeal or delay were only self-serving roadblocks to good policy-making.[61] The flavor of such an argument is conveyed in remarks by Gov. Bob Martinez following the legislature's preemptive action:

> The issue of how and where to dispose of Florida's waste is never an easy question to resolve, but the fact is that we have a thorough procedure in place to help us determine what is best for this state, for both the future

of our residents and the environment on which we all depend. As you remember, we did a thorough and costly evaluation of over 10,000 state-
owned parcels, using outside consultants who evaluated the parcels using
hydrogeologic, wetlands protection, air quality, transportation and other
environmental criteria. The decision was made after careful consideration,
using that process, and involved many public meetings around the state. I
believe we should all support that process now.[62]

The same sort of argument, strictly segregating the political from the technical
issues in siting, was used by Secretary Twachtmann:

There was absolutely no politics that I had any awareness about. We ran
carefully down that whole process trying to find which piece of land was
the best. We found five, and then the [ERC] selected the best one.[63]

These statements reveal the limits of the managerial approach when it is
applied to complex issues of social regulation. They ignore the degree to which
such decisions defy the simple segregation of political from technical issues and
the importance of fair process to the legitimation of any decision. First, it seems
clear that UCAP had raised legitimate questions about the consultant's report and its
use by DER. Although there might not have been anything sinister about it, it is
curious that the Union County site moved rapidly from near the middle of the
original rankings to the top in the final list. Moreover, many of the considerations
identified by Martinez, above all geographical location and transportation, would
seem to argue against such a remote site.[64] Once the ERC designated the site,
policymakers labeled the entire process a technical, objective, and nonpolitical
exercise that had identified the supposed best location, rather than addressing the
many anomalies that cropped up in the decision-making process.[65] Opponents of
the outcome could then be portrayed as self-serving NIMBY protesters, working
against the public interest.

Ironically, just as the legislature prepared to confirm the Union County site
for the HWDF as the best location in the state, a private firm began negotiating with
the DER to build the statewide HWDF in Polk County, in the south-central part of the
state. DER Secretary Twachtmann seemed enthusiastic about the prospect of a
privately sited HWDF and about the potential of the Polk County site:

I think I would probably prefer the private side doing this [in Polk
County]. It will probably solve a lot of problems. We would have to put a
lot of controls to make sure that whatever the private site did is the right
thing, that it is the kind of treatment we need, and that it would be large
enough to solve our problem. It can be done more quickly on the private
side. My impressions of the Polk County site are pretty good, because it's

all in vast acreage of mined-out phosphate lands, and there are not any neighbors in the near area.[66]

If such an appropriate site could be found by one private facility developer, perhaps the site search of only state-owned land required by the WQAA was simply too constrained to discover more appropriate parcels in Florida, a point often made by opponents of the Union County site.

Second, by assuming a technically correct decision, these arguments ignore the importance of procedural fairness in legitimating the outcome of a decision-making process. So, for example, the comments by Martinez and Twachtmann ignore the ways in which the legislature changed the process established by the WQAA and relied upon by the citizens' groups in planning their challenge. Indeed, once the legislature had intervened and established the streamlined, one-stop appeal for the HWDF in DER, the only recourse for UCAP following DER's decision was the governor and the cabinet, both on record as already supporting the Union County site!

In contrast to the managerial orientation of policymakers, citizen groups and legislative representatives from their districts employed a critique of the siting process rooted in pluralist and communitarian languages. Speaking past the policymakers quoted above, they focused on the legislature's violation of the policy-making process established by the WQAA and on the belief that Union County was chosen because of the poverty and powerlessness of its residents. This critique, pointing to the irregularities in the consultant's report to the ERC, argues that the siting issue was inevitably political and that politics, not expertise, could better explain the actions of the legislature. Responding to Governor Martinez's defense of the designation of Union County, Sally Gotts, president of UCAP, rooted her objections in the pluralist argument that process is important and needs to be protected. Abandoning agreed-upon procedures, regardless of the technical justification, was, for her, a violation of democratic norms and could not be defended:

> Now, they're [lawmakers] designating a site because it's not going their way. All of a sudden, we need to change the game, change the rules and do it another way. That is not democracy.[67]

Sen. Sherry Walker, whose district included Union County, made an argument even more pointedly rooted in the pluralist rhetoric:

> I am very disturbed by an attempt to circumvent the Union County folks' right to have a hearing. They will never accept it [the site], but they won't feel as railroaded with the whole thing if they have a hearing.[68]

This language recognizes the importance of fair treatment. At a minimum, the decision-making process should have conformed to democratic norms because process as well as outcome legitimates policy-making.

Senator Walker, in accounting for the legislature's behavior, connected this argument, at least implicitly, with a rejection of the managerial approach, which defines the issue as a technical one. Legislators voted for the Union County site not because it was technically the best site, she argued, but because their vote would prevent a facility from being located in their own districts:

> This will definitely be a not-in-my-back-yard issue. I don't think there is any senator that's going to stand up and say they want it in their county.[69]

But her analysis also invoked the communitarian language, as she emphasized the inequity of her constituents' battle against the sophisticated forces in favor of the Union County site. She mentioned the struggle it had been for the poor citizens of Union County to raise the money to hire the attorneys and experts needed to analyze the information relevant to the siting decision and to challenge the process in the administrative hearing:

> They've sold too many chicken dinners and put together too many of their nickels and dimes to hire a lawyer for us to just take away that hearing. I just want them to have their day in court.[70]

In Polk County, which emerged almost out of the blue as a possible site for the HWDF the day after the legislature designated Union County, local citizens were even more pointed in their belief that their county was chosen because of their political powerlessness. The following remarks of three Polk County residents contain a poignant rejection of the pluralist perspective, which assumes the adequacy of the existing structures of representation. Although not yet as organized or articulate as the spokespersons for the Ironbound Community in New Jersey, these residents voice the beginnings of a communitarian rejection of conventional policy-making:

> We've got so many things against us. The water is completely contaminated, and they're [also] wanting a high power transmission line right through Bradley [in Polk County].

> The area has more or less been a dumping ground for central Florida for years. The area looks like it's been bombed out.

> We live in the country, but we are here. I'm very concerned about it. It makes me feel like they're just crossing us off—like we're not here.[71]

The crux of this debate between citizens and policymakers is neither that the citizens have no capacity to understand HWDF siting and therefore no business challenging technical expertise, nor that the policymakers are simply trying to force through an unsubstantiated decision in ways that violate democratic processes and trample the interests of the poor. Rather, it is the failure of policymakers to understand the political dimensions of such decisions and the degree to which experts alone cannot legitimate HWDF siting. Given the focus of the WQAA on education and awareness, it is ironic that no one anticipated the reaction of this newly aware citizenry to unpopular siting decisions. UCAP and citizens' groups like them had acquired the ability both to understand the technical issues and to critique, on reasonable grounds, the assumptions made by other experts, for example, in their challenge of the Westin, Inc., study. Florida's policymakers seem to have paid little attention to how informed citizens might have been incorporated into the policy process. At the very least, it seems clear that for the siting decision to be legitimated, UCAP needed its day in court to air its challenge of the DER decision. The precedent established by the state in violating an established appeals procedures, and thus eliminating the opportunity to voice disagreement in a meaningful institutional setting, is likely to make citizens' groups much more distrustful in the future. Regardless of where the HWDF is built (if it ever is), the entire episode will make the next attempt at making redistributive regulatory decisions—especially if they depend upon challenged technical information— even more conflictual and difficult.

Our arguments suggest that ongoing participation is a prerequisite for the development of the trust and shared language required to allow political systems to address the inevitably conflictual issues that are part of social regulation. Instead, Florida's officials assumed the acquiescence of an informed citizenry. This was a fundamental error, based upon a mistaken assumption that the regulatory decisions were all technical, that all experts would agree, and that politics could somehow be kept out of the decision-making process. Omitted from the process was any sustained attempt to draw the citizenry into policy-making or to resolve the political differences between informed and aroused citizens and regulatory officials.

As it turned out, citizens could participate in only the most contentious aspect of the whole siting process. They were informed virtually after the fact about the ranking of their county as the site and subsequently dispossessed of the normal opportunity to appeal by a legislature intent on making the site decision stick[72] regardless of its technical underpinnings. In such hostile environs, with all parties arguing in entirely different languages of regulatory legitimacy, compromise and accommodation are the least likely results.

Conclusions

Each of our case studies illustrates the difficulties that states and localities have in managing their hazardous waste problems. In all three states, we found that when people disagreed about hazardous waste policy, they consistently expressed themselves in terms of the competing languages of regulatory legitimacy. Often in the debate over HWDF siting, managerial rhetoric encountered pluralist discourse. However, local citizens, faced with the possibility of having to live near hazardous waste sites, spoke with a communitarian voice. Beyond the bounds of either managerial or pluralist discourse, this voice gave expression to their anger and resentment over being powerless vis-à-vis government bureaucracy and private industry. In all three states, political institutions were unable to resolve or even address the issues raised by the communitarian challenge.

No state could resolve the central issue raised by the dialogic model: creating and empowering citizens capable of dealing constructively with the redistributive consequences of regulatory policy in an advanced industrial democracy. The states' lack of experience in dealing with such profound issues in the context of concrete and specific policy problems doomed their efforts. In Florida, following the arousal of public concern with hazardous waste, the state chose to ignore the prospects for informed debate by preempting the process when its results were challenged by informed citizens. In Ohio, siting was discredited as the HWFB's disingenuous attempts to hurry the siting process along took advantage of local groups too poorly organized to protest or even participate in the board's decisions. But the advantage was only temporary, as citizens mobilized to challenge the regulatory regime. In New Jersey, the painstakingly careful process fell apart as the state's strategy for group representation produced irreconcilable splits between state-level environmental groups and the locals likely to be affected by siting decisions. Their exclusion led grassroots groups to challenge the criteria for site selection.

The failures of federal and state regulation of hazardous wastes mobilized citizens like the ones discussed in our case studies, and they began addressing their policy options in ways not recognized by conventional politics rooted in either managerial or pluralist languages. Their presence disrupted the entire policy process from the ground up. Their critics discredited them with the NIMBY label and called them small-minded and selfish—an obstacle to responsible hazardous waste management. But grassroots groups offered an alternative to the policy gridlock of more conventional approaches by challenging the traditional relations between market and government, democracy and capital, in ways that only the communitarian language can express.

This chapter documents the centrality of citizen protest to the issue of hazardous waste management. But the nature of that protest is not well understood, precisely because it is politicized within the terms of communitarian discourse. In the next chapter we examine the emergence and articulation of the grassroots protest against toxics policy as a communitarian phenomenon, mindful of the limitations that have traditionally kept the democratic wish from transforming direct democracy into policy influence. For the communitarian voice to be heard in a true policy dialogue, the protests of citizens at the grassroots must transcend the NIMBY label. Chapter 7 considers whether they have and how they speak to the reform of social regulation in America.

7 / Not-in-My-Back-Yard, Right to Know, and Grassroots Mobilization

So far, our analysis of hazardous waste regulation reveals how contradictory languages of legitimate policy-making compete with one another in the context of American federalism. In the early 1980s, national policymakers exploited both the federal system and the rhetoric of regulatory legitimacy to distort the implementation of hazardous waste policy. At the local level, the effects of this distortion were magnified as state governments and communities wrestled with the redistributive consequences of hazardous waste policy, again using alternative discourses of policy-making in ways that could breed only misunderstanding and resentment among all parties. The end of the decade witnessed a pervasive dissatisfaction with the regulatory process and policy gridlock at the grassroots, as communities expressed their sentiments through the now-familiar Not-in-My-Back-Yard (NIMBY) protests whenever and wherever HWDF siting became a local possibility.

Our dialogic critique has emphasized the incompatibility of the managerial, pluralist, and communitarian understandings of social regulation in general and of hazardous waste regulation in particular. Each language conveys an incomplete characterization of a linchpin of that policy's failure—local opposition to facility siting. We argue further in this chapter that the more familiar managerial and pluralist approaches miss the point about NIMBY, and the less-recognized communitarian perspective, while flawed in its own right, casts the phenomenon in a revealing and more favorable light. Neither the managerial nor the pluralist language can account adequately for the widespread, enduring vigor with which grassroots opposition has halted HWDF siting and ultimately affected hazardous waste policy in general. Managerial appeals to reason and pluralist references to process miss the mark targeted by the so-called NIMBY protesters and invariably lead to negative characterizations of the phenomenon.[1]

From a managerial perspective, HWDF protesters are simply incapable of appreciating the information they are being given about hazardous waste cleanup and the safety of disposal facilities. Managerial critics often find fault with local siting opponents for failing to compare the "relative risks" of "proper" hazardous waste disposal with, for example, those of ozone depletion or even those of radon in their own homes. This view characterizes popular resistance as either naively ignorant or cynically self-interested. The typical remedy suggested here is the injection of greater expertise into the policy process, either through increased grants of authority to experts or better citizen education, so that the affected public might more rationally evaluate the risks they face with hazardous wastes or at least respect more the expertise of professional decision makers. If better analysis fails to reduce community resistance, then the managerial approach generally assumes that government can calculate a level of economic compensation that will negate local opposition and render policies acceptable.[2]

The pluralist critique of NIMBY, with its emphasis on interest representation within a fair decision process, stresses the unwillingness of grassroots opponents to accept decisions that have been consummated "democratically."[3] Pluralists argue that, regardless of the risks involved, a constitutionally conceived process providing for adequate group representation should render hazardous waste policy decisions acceptable even to those who do not narrowly benefit from the results. Groups willing to benefit from the political process, so the argument goes, must also be willing to accept unfavorable outcomes in order for that process to serve the public interest. Given its paradigmatic emphasis on compromise, the pluralist remedy for NIMBY generally revolves around some form of environmental mediation in which representatives of the different interests negotiate an appropriate site and settlement.[4]

The communitarian understanding of NIMBY differs fundamentally from these managerial and pluralist assessments and accounts for NIMBY's political impact in ways that neither of the other languages can explain. First, the other languages fail to anticipate NIMBY's raw success in stalling hazardous waste policy implementation. Second, they underestimate the remarkable mobilizing force that the NIMBY phenomenon has brought to local politics—its capacity to inspire and maintain direct citizen participation.[5]

These results are inexplicable in managerial and pluralist terms because neither the managerial emphasis on expertise nor the pluralist focus on process appreciates the role that direct citizen participation has in legitimating or challenging policy outcomes. In contrast, such direct participation is central to the communitarian understanding of legitimate policy-making.

Against the negative characterizations typical of managerial and pluralist commentary, the communitarian perspective constructs NIMBY in a positive light. It locates NIMBY within that long-standing protest tradition of American democratic politics that seeks, however vainly, citizen self-government in the determination of public policy outcomes.[6] Essentially, the communitarian language invokes values that are unrepresented in the other two languages and that seek expression when the operation of market or government (or both) crosses unrecognized barriers of public acceptance. In this light, NIMBY signifies such an occurrence.[7]

NIMBY and the Evolution of the Communitarian Discourse

Because the communitarian language addresses issues of political cooperation and conflict at the local level, it is relevant precisely at the point where hazardous waste regulation fails. And our dialogic model suggests that ignoring the communitarian language hinders the search for remedies for the problems of social regulation. But communitarian discourse is as incomplete in its conception of politics and policy as are the managerial and pluralist languages. Thus, resolving the policy gridlock surrounding hazardous waste exclusively in communitarian terms also is problematic.

Specifically, in chapter 3 we raised three related questions about the communitarian interpretation of social regulation. These are a central focus in this chapter, as we evaluate empirically the role of the communitarian perspective in hazardous waste regulation and its potential contribution to a dialogic approach to social regulation. First, we suggested that because the communitarian language is bound to an understanding of local political reality, it fails to comprehend the societywide impact of community-based action. This limitation of the communitarian language is, of course, the foundation of the managerial and pluralist criticism of NIMBY used to discredit HWDF protesters. It raises a very important empirical question about communitarian-inspired political action: Can local protesters, articulating their demands with a communitarian vocabulary, develop a national agenda for dealing with hazardous substances? Or does their vision of policy stretch only as far as their own backyards?

A second weakness of communitarian discourse is its grounding in traditional values, which resist progress and the consequences of modernity with evocations of family, religion, and, of course, community. A nostalgic vision of politics makes the communitarian language ill-suited for appreciating the essential scientific and technical dimensions of hazardous substances policy, which must be included in a dialogic approach to social regulation. How do NIMBY groups address the issue of scientific and technical information—and the expertise to interpret it—within a communitarian framework so suspicious of these essential elements of social regulation?

Third, the communitarian language emphasizes local political mobilization with its rhetoric of citizen self-government and direct democracy; but, as James Morone illustrates, the democratic wish has consistently failed to sustain itself institutionally.[8] Grassroots activists can block HWDF siting by overwhelming local governments, but what institutional means can preserve their agenda over the long haul in a complex system of federalism? Are they doomed to a local existence, enduring only so long as a toxic threat remains imminent? The answers to the first two questions obviously depend upon the answer to the third, since the institutional sustenance of the grassroots agenda is a precondition for bringing communitarian discourse to bear on hazardous substances regulation as part of a dialogic policy process.

NIMBY, Information, and Empowerment

As a prelude to our empirical examination of the NIMBY phenomenon, we examine exactly what the term *NIMBY* implies and whether or not it accurately captures local resistance to HWDF siting, for which it has become such a popular label. A better appreciation of the communitarian interpretation of NIMBY allows one a more balanced grasp and strips away the pejorative connotations associated with managerial and pluralist uses of the term. In the process, we can begin to analyze the three limitations of the communitarian language discussed above.

The communitarian language describes NIMBY as a bid for empowerment at the local level. But exactly how does NIMBY become a source of empowerment from a communitarian perspective? The answer lies in the communitarian understanding of the connection between the initial motivation of local citizens to protest the siting of HWDFs and their resulting exposure to information about the daily hazards to which they and their families might be exposed. Because of the obvious risks to the affected community, the proposal of an HWDF site hits close to home and typically mobilizes citizens like no other political event. Whether they are inspired by a concern for property values, the water supply, or future generations, residents tend to get involved in siting controversies. The communitarian language interprets this involvement as a community's political struggle to control its own fate.

Accordingly, their struggle reveals to residents how industrial practices and government policies, long ignored or veiled in secrecy, introduce risks to community safety. In communitarian terms, protesting HWDF siting is a form of consciousness raising, and the lessons learned from their protest impress upon nascent citizen-activists the connection between information about the risks they live with and, lacking access to that information, their powerlessness to affect those risks.

This communitarian appreciation of information complements our emphasis on the crucial role information plays in the government's definition of the boundaries between democracy and capitalism in our society (see chapter 2). The regulatory definition of those boundaries is judged by two conventional standards: efficiency (the managerial benchmark) and fairness (the pluralist criterion). To review our argument, social regulation suffers on both counts for reasons related to information. On the one hand, much of the information necessary to formulate social regulatory policy is controlled by experts in government and private industry. While expertise is essential for arriving at regulatory solutions, the technical nature and location of both information and expertise violate the pluralist standard of fairness by preventing those who must bear the consequences from being effectively represented in the regulatory process. On the other hand, ironically, information about the risks, costs, and benefits of social regulation is often uncertain enough or sufficiently contested to render its interpretation by experts arbitrary, thus calling into question the managerial efficiency of regulatory solutions.

This peculiar combination of policy problems befuddles conventional discourse about social regulation. The managerial language is trapped in a sort of infinite regress in which the challenge to expertise is met by relying upon even more experts, while the pluralist language addresses the problem of access by resorting to elite representation of affected interests or, worse, by assuming that token participation in the policy process will satisfy grassroots constituencies. Neither comes to grips with the issue underlying this regulatory dilemma: the generation and command of information.

In contrast, communitarian discourse addresses the information problem by redrawing the boundaries among market, government, and community. It envisions information as a public right rather than a private or privileged commodity. No longer should private industry be permitted to control information about practices that might risk harm to the surrounding community. Neither should such information be selectively entrusted to government experts. Instead, affected citizens—workers and residents—should be permitted direct, unfettered access to relevant information, so that they can begin to command the knowledge necessary to take part in community life, which in this case means constructing a hazardous substances policy.[9] By challenging the conventional assumption that information is private property, the communitarian language embraces the old adage that "knowledge [commanding information] is power." The essential component of meaningful participation in the social regulatory process is this command of information, and in broadening access to this command, communitarian discourse is empowering because it permits the expression of an alternative conception of legitimate regulation, a conception championed by those at the grassroots who feel excluded from the policy-making process.[10]

Whither NIMBY?

The first limitation of the communitarian language—its local bias—raises the empirical question of whether NIMBY is really an accurate characterization of the activities of local protesters. The NIMBY label itself conveys perfectly the local bias criticism. As applied by critics of HWDF siting protests, it develops the following scenario: Activists arise in communities targeted for HWDFs and generate "campaigns of fear"[11] to mobilize sufficient local opposition to defeat the proposed site.

The HWDF proposal then moves to another site, while the mobilized community returns to normal, without a thought about the larger consequences of their opposition for the growing hazardous waste disposal crisis.

The NIMBY label thus implies that the protesters' principles extend only to their community's boundaries. This strategy is irresponsible, its critics contend: successful enough to defeat HWDF siting without suggesting an alternative policy in its place. Such a characterization would surely cast doubt upon the communitarian understanding of resistance to HWDFs. An examination of the origins of the NIMBY label sheds light upon the accuracy of its use here and, at the same time, addresses our concerns about the first limitation of communitarian discourse.

References to the term *NIMBY* emerged in the early 1980s, but the phenomenon of community opposition to "locally undesirable land uses," or LULUs, is at least as old as the passage of zoning ordinances,[12] which have been used traditionally to separate industrial, commercial, and residential land uses. Municipal zoning ordinances are, of course, an important tool that local governments use positively to plan the growth of their communities, but these ordinances can also be employed negatively to exclude land uses vital to metropolitan areas.

The exclusionary aspect of zoning has always been a challenge for municipal governments, particularly as cities became crowded and as the areas around them grew into self-governing suburban communities.[13] Urban jurisdictions found it increasingly difficult to pursue their land-use agendas, including sanitation, corrections, and public housing initiatives, which had traditionally been built on the outskirts of town. Such was the context within which the term *NIMBY* emerged, and it was first used as a pejorative to describe the efforts of well-to-do suburban homeowners to exclude, among other things, the development of moderate- and low-income housing in their communities.[14]

Given this background, how appropriate is the NIMBY label for those protesting the siting of HWDFs? First, it should be remembered that the term has long been used to impugn the motives of those who object to a wide variety of development schemes, regardless of the public virtue of those schemes. As recently as 1991, a blue-ribbon commission assembled by U.S. Department of Housing and Urban Development Secretary Jack Kemp and chaired by Thomas Kean, Republican former governor of New Jersey, issued a report entitled "Not In My Back Yard: Removing Barriers to Affordable Housing."[15] The 150-page report is highly critical of community efforts to control residential expansion. On the surface (and, of course, prominently in the title), this report targets the selfish interests of residents as a barrier to the construction of more affordable housing. The following passage illustrates the tone of the report: "The heart of NIMBY lies in fear of change in either the physical environment or population composition of a community. Concerns about upholding property values, preserving community characteristics, maintaining service levels, and reducing fiscal impacts are often involved. Sometimes these expressed concerns are also used as socially acceptable excuses for ethnic and racial prejudices. Whether genuine or used as excuses for other motives, such concerns often generate strict development curbs."[16]

In fact, the report's main purpose is to advocate "streamlined" local permitting processes so that developers would be able to avoid "delays" caused by public input, local building codes, and environmental regulations, as if such codes and regulations were motivated solely by the NIMBY attitudes of residents and had no relation to construction quality or environmentally sound development. Not only does this report represent a politically motivated use of the NIMBY label, but it also indicates essential differences between the narrow NIMBY attitude reflected in suburban housing disputes and the expressed attitudes of those protesting HWDFs at the grassroots. Nowhere in the above quote are the themes of environmental protection, safety, and health mentioned; yet, as we repeatedly document in chapter 6, these concerns are central to citizens opposed to HWDF siting.

That the NIMBY label might be applied overbroadly to discredit local opposition to HWDF siting draws attention to a second issue regarding the appropriateness of the NIMBY label. Results from a survey of citizen attitudes toward HWDF siting, reported by Kent Portney, suggest a demographic split between those who object to HWDFs across the board and those who are not opposed to HWDFs in general but reject the siting of HWDFs in their own communities. The latter group tend disproportionately to have high incomes, to be educated, and to have a Republican party identification.[17] Portney refers to these appropriately as "NIMBY Syndromers," and they seem very much like those suburbanites described empirically in many works documenting the gradual changes in the modern metropolitan political landscape.[18] But if the suburban political profile captures Portney's NIMBY Syndromers, it certainly does not describe the profile of either the grassroots activists we described in chapter 6 or those we discuss below.

In fact, Portney's analysis draws attention to this distinction. He divided his survey of Massachusetts respondents into three groups: those who favored siting in the state and in their community; those who were opposed across the board; and the NIMBY Syndromers mentioned above. A profile of the second group reveals a middle- to lower-income respondent overwhelmingly concerned about the health risks associated with hazardous waste disposal, likely to be a female, and specifically a mother with children.[19] In a separate national survey using the same groupings, Portney found that those in the second group were much more likely to read a newspaper every day and watch nightly television news broadcasts than either the pro-HWDF group or the NIMBY Syndromers.[20]

This evidence indicates that there are demographic differences between those who oppose HWDFs on principle and those who object to HWDF siting only when it is nearby. It appears, then, that opposition to HWDFs is more complicated than the NIMBY label implies. In fact, a substantial majority of those who objected to HWDF siting in Portney's surveys did not fit his characterization of NIMBY Syndromers.[21] These findings raise the question of whether opposition to HWDF siting transcends the exclusionary impulse associated with suburban opposition to metropolitan land-use proposals. The implications of these findings for evaluating local opposition to HWDFs bode well for the possibility of transcending the limitations of communitarian discourse in a dialogic reform of social regulation. As noted above, it is politically expedient to discredit HWDF opponents by labeling them as nothing more than selfish suburbanites or ignorant naysayers, but the empirical evidence, though not conclusive, casts doubt on such characterizations.

These doubts are confirmed when one looks at the history of HWDF opposition. Its origins and evolution are drawn from a context far removed from the suburban phenomenon of NIMBY protests. The roots of siting opposition are grounded in the larger context of grassroots citizen activism in the 1970s. As we address that history below, evidence regarding the relation between it and the second limitation of the communitarian language—the aversion to scientific and technical information and expertise—emerges.

Origins of the Movement against Toxics

In 1978, Janice Perlman assessed the state of grassroots organization around the country and noted that the decade had witnessed a mushrooming of "citizen action" groups. She cited surveys indicating that there were somewhere between four thousand and eight thousand neighborhood groups then active in America.[22] Perlman created a detailed typology that divides these groups into two categories: "issue-oriented / direct action" groups and "self-help / alternative institution" organizations. Groups in the first category have roots in the political activism of the late 1960s, contain a populist mix of workers, neighbors, and ethnic groups, and are committed to one or more of a variety of issues ranging from tenants' rights to environmental quality. Those in the second category typically come from the earlier tradition of civil rights protests and often are composed of ethnic minorities with a focus on service provision or economic development.

Although the number of groups expanded greatly in the 1970s, Perlman's typology suggests that their existence was tenuous, especially among those of the issue-oriented / direct-action type; a typical group's life span might be no more than five years. Such data provide evidence supporting the third limitation of the communitarian language—the staying power of community groups. But Perlman also noted that some groups survived the five-year watershed. The surviving groups were likely to have secured consistent funding and maintained a full-time, professional staff, but, just as important, they were typically able to connect effective local organizing to a national agenda, in contradiction to the communitarian language's local orientation. Many groups in the late 1970s began forming coalitions with other grassroots organizations, public interest groups, and labor organizations, a practice that proved vital for local opposition to HWDF siting.[23]

The National Context   Perlman's analysis of grassroots activism accurately portrays the local political ferment within which key elements of the movement against toxics emerged. Local mobilization followed a gradual, decade-long course of government recognition of the problems posed by hazardous substances. This paralleled a more general public concern about toxics in areas of public health and environmental degradation.[24] For example, we mentioned in chapter 2 the still-controversial Delaney Clause, amending the Food, Drug, and Cosmetic Act.[25] Congress passed this amendment in 1958, signaling government's preoccupation with cancer-causing food additives even before 1960. Soon thereafter, public awareness over toxic chemicals in the environment was piqued by Rachel Carson's landmark study *Silent Spring,* published in 1962.[26] Concern about toxics continued to grow during the remainder of the 1960s, but it found legislative expression at the national level only after the more general issues of air and water pollution and occupational safety and health had been addressed.

We have already related the second-generation challenge that hazardous substances posed for the new social regulation. The public sentiment and political entrepreneurship that produced Earth Day in 1970 had inspired first-generation environmental legislation, and this established the regulatory apparatus necessary to bring to light the extent of the toxics problem.[27] In particular, the regulations and research generated by EPA laboratories and OSHA—through NIOSH—began to document an overwhelming number of potentially hazardous substances facing residents and workers across America.[28]

Armed with evidence of the large number and broad distribution of toxics in the environment, health and environmental lobbyists in Washington spurred congressional investigation of public and private sector practices in the testing, handling, and disposal of toxics. But we pointed out in chapter 5 that increasing governmental and public awareness about toxics came at a time when congressional entrepreneurs, federal agencies, and environmental lobbyists were finding it difficult indeed to maintain the regulatory momentum with which the decade had begun. Recession dogged the national economy in the mid-1970s, and conservative critics of regulation were attacking the fragile environmental coalition in Congress by pitting the issue of unemployment against the increasing cost of additional environmental and workplace regulation. Despite these obstacles, the two legislative mainstays of hazardous substance regulation—RCRA and TSCA—became law in 1976.

EPA's delays in implementing the 1976 toxics laws[29] and OSHA's continuing problems in regulating hazardous substances in the workplace[30] ironically paralleled extraordinarily productive efforts by both agencies in cataloging potentially

toxic substances. In 1977, EPA and OSHA, along with the Food and Drug Administration and the Consumer Product Safety Commission, addressed this irony by
forming the Interagency Regulatory Liaison Group to coordinate a so-called
Cancer Policy. The Cancer Policy was designed to be a systematic means of
interpreting scientific evidence about carcinogens, ostensibly to streamline the
process of identifying hazardous substances and regulating their presence in a
variety of contexts.[31]

But economic decline, industry challenges, and agency delays in implementation combined to produce an impression that the federal government's regulatory
efforts were costing too much and accomplishing too little in regulating hazardous
substances. An emerging disenchantment with federal social regulation put environmental groups based in Washington, D.C., on the defensive as well. Opinion
polls in the late 1970s indicated that public support for environmental values was
slipping.[32] By the end of the decade, conservative critics were labeling environmentalists as elitists, willing to sacrifice workers' livelihoods for questionable
environmental regulations.[33] Meanwhile, media coverage of an increasing array of
toxic substances and their presence in repeated exposures, spills, and accidents to
workers and residents in communities across America produced dramatic examples of the dangers associated with industrial, agricultural, and military handling
of toxics.[34]

Committees on Occupational Safety and Health: The Local Reaction   As the national leadership of labor and environmental lobbies worked to close rifts over the
trade-off between jobs and environmental regulation, local union and citizen
activists were becoming increasingly alarmed about industrial practices in their
communities. Their concern was fueled by the accumulating data on toxics, mentioned above, and particularly by research results from NIOSH.[35] But local activists
criticized the federal government for failing to respond effectively to the problems
it was documenting. This irony only reinforced doubts about the priorities of
national environmental and labor elites, thus inspiring the formation of the grassroots toxics movement, a precursor to later developments in the local opposition to
hazardous substances policy.

The movement coalesced from sources as disparate as public health officials,
local union activists, and community organizers, all of whom had become active
independently in the early 1970s, during the period of intensive grassroots organizing described by Perlman. The first coalition groups emerged in large northeastern and midwestern urban centers and usually called themselves Committees
(or Coalitions or Projects) on Occupational Safety and Health (COSHs). The political agendas of COSHs varied widely, depending upon local conditions. Taken together, their disparate agendas reflected local activists' ambiguity not only toward
union leadership and environmental elites, but also toward OSHA as it grappled with

workplace health and safety during its first decade. Some COSHs mobilized specifically to challenge OSHA's lackluster implementation of its congressional mandate. Others worked more closely with OSHA, and, in the late 1970s, many even received organizing grants from OSHA to foster employer/employee education programs.[36]

The unifying features of these groups were their concern over the accumulating evidence of occupational disease caused by workplace and community exposures to production hazards known only to employers and their commitment to reforming the workplace through worker education and increased worker participation in workplace management, with an emphasis on health and safety.[37] The new militancy of COSHs contrasted sharply with conventional labor politics at the local level, which had long since abandoned control of the workplace to management and had distanced itself from community political involvement as well.[38] Directly challenging traditional union tactics, many of the successful COSHs openly embraced community activists in order to pursue their reform agendas.

The COSH movement is relevant to our ideas about the limitations of the communitarian language. The origins and advocacy of COSHs appear to be in most respects communitarian, yet they also embraced values that transcend the limitations of the communitarian language, especially its parochial bias and aversion to complex scientific and technical information. The coalitional nature of COSH mobilization brought labor out of the workplace and community organizers out of the neighborhood to join forces locally (and to network nationally, as discussed below) over precisely the subject matter that communitarian discourse traditionally avoids. In fact, they were devoted to providing workers and residents with the critical capacity to understand scientific and technical information about toxics in the workplace and the community so that they could make informed choices about their livelihoods. Thus, the very nature of the COSH movement during the 1970s seems to have expanded the horizons of the communitarian language. By the end of that decade the movement's right-to-know campaign symbolized that expansion.

The Right to Know   COSHs were well established at the grassroots when the revelations about hazardous wastes buried at Love Canal surfaced in 1978, and the timing could not have been more fortuitous for the success of the grassroots movement against toxics. Because many COSHs already pursued a toxics agenda and included community activists in their ranks, they became a focal point in the grassroots response to hazardous waste, thus further encouraging COSHs to broaden their scope beyond the workplace into the community. But in the late 1970s information about the location and contents of HWDFs around the country was still far from comprehensive. Activities inspired by leaking landfills and toxic spills nearby remained sporadic, driven by the isolated circumstances of each incident.

The catalyst for a coherent grassroots strategy for combating toxics at the local level was extensive media coverage in 1979–80 of elevated cancer-related death rates in the industrial centers of the Northeast and Midwest. Epidemiological studies revealed that certain cities, particularly Philadelphia and Cincinnati, had comparatively high cancer rates. Suspicious that these rates were related to industrial exposures, grassroots groups, first in Philadelphia, came together to address the issue. Susan Hadden notes the significance of this crucial meeting:

> In 1979, the Philadelphia occupational safety coalition PHILAPOSH [Philadelphia Area Project on Occupational Safety and Health] and the Environmental Cancer Prevention Center sponsored a conference on toxic substances in the workplace. In the course of the discussions, attendees asked why only workers should know about hazardous substances; all members of the community are exposed to the same risks, albeit at lower intensity, by permitted and accidental emissions, as residues on workers' clothes, and through the passage of vehicles transporting the materials. Thus was born community right to know.[39]

The notion of right to know—defined as worker and community efforts to gain information from industry[40] and government about the nature of hazards (particularly chemical and radioactive) under their control—has its historical and philosophical roots in the communitarian impulse, discussed throughout this book, that undergirds freedom of speech and the press in the U.S. Constitution's Bill of Rights. It is no coincidence that the first use of the term can be traced back to James Wilson, the advocate for strong democracy at the Constitutional Convention (see chapter 2). He was an outspoken advocate of the people's "right to know what their agents [in government] . . . have done."[41] Wilson's views on the importance of public access to information were integral to his understanding of citizenship and point the way to appreciating how access to information can empower people as citizens: first, because of the centrality of access to information as a path to political empowerment (mentioned earlier in this chapter) and, second, because of the effect that this access has on reshaping the Madisonian relation between private property and public power.[42]

The right to know was the vehicle the grassroots movement against toxics used not only to redefine power in the workplace, but also to foster a new sense of citizenship among its participants, a sense alien in many respects to the NIMBY characterization fostered by the movement's critics. In discussing the rise of the right-to-know movement and its integration into the opposition against HWDF siting, we emphasize the language used by participants in the movement to describe and defend their actions. We include their comments not because we necessarily agree with them, but rather to expose the reader to the communitarian features of their discourse and to their struggle to overcome the limitations of that vocabulary as they defined and redefined politics, property, and regulation.

The formative experiences of the groups that met in Philadelphia in 1979 bear emphasis. First, although the initiative for an employee right to know preceded community right to know[43] by several years, it involved some of the same people who attended the Philadelphia meeting. The current idea of right to know started as a trade union goal in 1975, when several COSH groups received private funding for a grassroots campaign for a federal chemical labeling standard from OSHA.[44] A year later, assisted by Ralph Nader's Health Research Group, PHILAPOSH formally petitioned OSHA for the standard. But OSHA, under intense pressure from industry, temporized for more than four years in issuing the labeling standard. Out of frustration, in 1979 the Philadelphia meeting was called in the hope that a local right-to-know initiative might prompt OSHA to act.

That meeting began a campaign that led in 1981 to a pathbreaking set of local ordinances affording residents, local officials, and laborers access to industrial toxics information. Following Philadelphia's example, COSH groups and citizen activists in other states and localities followed suit. For example, in 1982, a coalition of the Ohio River Valley Committee on Occupational Safety and Health, the Ohio Public Interest Campaign, and the United Auto Workers local engineered passage of a right-to-know ordinance in Cincinnati, a city with the highest cancer death rate on record at that time. Within four years of right to know's passage in Philadelphia, there were at least twenty state laws and forty municipal ordinances containing a variety of right-to-know provisions.[45]

Industry's reaction to the spread of right-to-know legislation across the nation followed two primary lines of objection: first, that right to know involves needless paperwork and expense given the benefits produced; and, second, that revelation of hazardous substances in production processes would jeopardize firms' trade secrets. Underlying these objections was the much larger issue of how worker access to information about the production process might threaten the balance of power in the workplace.[46] The bid for knowledge about workplace hazards can be seen as a challenge to the old agreement between labor and management about control of the workplace (see chapter 4). It raises the prospect of worker participation in determining plant safety. Charles Noble describes the managerial tone of management's attitude:

> Taken seriously, worker participation conflicts with managerial control
> over the labor process. Effective participation means that workers exercise a
> certain degree of autonomy at work and cooperate among themselves. Both
> facilitate employee resistance to managerial directives. Therefore, employ-
> ers who seek to maintain and augment their control are likely to develop
> highly bureaucratized and centralized forms of personnel management.
> They will divide production processes into highly differentiated job struc-
> tures, rotate workers among different jobs, and limit the time and

opportunity available to workers to form affinity groups that might provide the basis for worker opposition to managerial prerogatives over plant administration issues. They will view challenges to the decisions and practices of plant administrators as challenges to managerial control of labor practices.[47]

Because the grassroots coalition was a potent local political force, industry was justifiably preoccupied that, in complying with right-to-know laws, they would be forced to reveal the nature of risks associated with doing business in the community, thus destabilizing relations among the private sector, labor, local government, and the community and threatening industry's freedom to continue, or to introduce unannounced, risky practices in their production processes. Wherever state and local right-to-know initiatives were proposed, industry resisted.[48] Indeed, industry resistance was so determined that it revealed important aspects of the capacity of federalism to contest grassroots initiatives, a point we return to later in this chapter.

In summary, the significance of the right-to-know movement for our purposes lies in its use of the persuasiveness of the communitarian language to mobilize citizens and, at the same time, its transcendence of some of the limitations of communitarian discourse. The movement's strategy had three key ingredients:

First, and most important, was its ability to access and use information about hazardous substances to force industry and government to reconsider conventional practices involving toxics. Access to this information enabled the movement to mobilize local residents and laborers anxious about the risks presented by hazardous substances in their communities and workplaces. Rather than avoiding scientific and technical information, the movement used precisely this information to educate and empower workers and citizens, thus guiding the communitarian impulse.

Second was the right-to-know movement's ability to link local groups across the country, so that a consistent front was erected against industry resistance to its initiatives. The movement purposely maintained a national network of similar groups such that locally developed initiatives could call upon a national constituency for support in the face of industry challenges. This feature challenges, at least partially, the local bias of communitarian discourse and sets the stage for the elaboration of a national grassroots agenda for dealing with hazardous substances, as discussed below.

Third was the movement's grassroots location within the structure of federalism. The tradition-breaking grassroots coalition between workers and residents combined with local government's reluctance to absorb the public burdens of risky industrial practices and made the movement's local political clout all the more irresistible.[49] Our discussion of local governments' inclination to resist redistributive policies (see chapter 4) helps to explain the power of the right-to-know movement at the local level as well as industry's and government's reaction, points we expand upon below.

"Not NIMBY, but NIABY"

The following remark by Caron Chess of the Delaware Valley Toxics Coalition captures the communitarian sentiment of those who pioneered local right-to-know legislation:

> Workers are the first victims of toxics exposure, then it's the rest of us. . . . Right-to-know is a unique opportunity for labor and environmentalists to work together. But right-to-know is not an endpoint—it's a beginning.[50]

So far, we have focused primarily on labor- and activist-inspired moves to gain access to knowledge about hazardous industrial processes. But more central to our subject is understanding how this connected to grassroots activism against HWDF siting. Are these groups capable of transcending their narrow concerns about hazards in their particular communities? To answer this question, we turn to the citizen side of the grassroots movement against toxics, examining the rise to prominence of one of the dominant grassroots organizations in the resistance to hazardous waste dumping—the Citizens' Clearinghouse for Hazardous Wastes, Inc. (CCHW).

Given our discussion in chapter 4 and the case studies we covered in chapter 6, it should come as no surprise that community activists could exploit the structural limits of local politics in order to thwart the redistributive decisions necessary for HWDF siting to take place. Local veto is something that even those local governments most sympathetic to HWDF siting eventually attempt, regardless of the state siting apparatus.[51] But a larger and more subtle issue remains because the protests engendered by the risks of hazardous substances have not remained community-specific, entirely isolated, or self-interested, and they have not simply faded away, as narrower interpretations might have predicted.[52]

What distinguished the citizen side of the grassroots toxics movement during the 1980s was its unconventional way of directing the political energy of residents newly mobilized by their discovery of a toxic threat to their community. As hazardous waste sites were found or accidents with hazardous substances occurred, activists in the affected communities typically organized groups to protest in ways only loosely associated with any established environmentalist strategies. Writing in early 1983, Ken Geiser describes the power of the appeal to community in that initial mobilization: "This new movement is bringing forth an environmental consciousness among people who were unlikely to think of themselves as 'environmentalists.' Because the movement is so tightly rooted in the immediate experience of people's community and family life, it has an urgency and a concreteness that is incredibly compelling. For these new 'environmentalists' environment is not an abstract concept. It is something which has already exposed them to hazards which are debilitating them and hastening their deaths."[53]

Geiser goes on to describe the typical three-stage progression of newly mobilized toxics groups, beginning with "Single Issue Protest Groups," which arose spontaneously and with the sole purpose of eliminating the hazard from their community: "The group is often informal or only loosely formed in structure, has a core group of leaders and a loose network of adherents who will come out for demonstrations or hearings. Tactics are developed spontaneously and range from letter writing to media attracting events that push the visibility of the situation."[54] The second stage is the "Multi-Issue Advocacy Group," often evolving from growing and maturing single-issue groups. Here, "awareness of region-wide or state-wide toxic chemical problems leads to a broader set of demands. . . . Tactics are linked to broader strategies, and particular events are staged to educate the public as well as capture media attention."[55] Multi-issue groups sometimes evolved into third-stage "Coalition Networks" linking local organizations together for general support and political leverage. . . . Neither government nor industry is seen as simple enemies. More typically the effort focuses on how to make government or industry more accountable or responsible."[56]

Geiser's interpretation of the grassroots movement's progression is a useful guide for reviewing the evolution of Lois Gibbs's Love Canal CCHW from a simple, spontaneous neighborhood campaign to an established, well-funded organization pursuing a consistent national strategy and committed to mobilizing and educating other groups at the grassroots level about the toxics problem.

The Love Canal incident inspired Lois Gibbs, a Love Canal homeowner and mother of two children, to launch a neighborhood organizing campaign that eventually forced the federal government to evacuate nine hundred families from that polluted neighborhood. Instead of using the relocation funds she received to buy another home, she set off for Washington, D.C., and used the money to found CCHW, one of the organizations responsible for the nationwide success of local HWDF opposition.

Other organizations, particularly the National Toxics Campaign, the Environmental Action Foundation, and the grassroots initiatives of Greenpeace, were also vital to this movement. We focus primarily upon the CCHW because it maintained a dedicated presence at the grassroots level of protest, while developing a unique national agenda. In this sense, then, the CCHW certainly represents the typical grassroots organization described earlier in this chapter by Perlman, but the CCHW also challenges that mold and, in so doing, sheds light on the prospects for including the communitarian agenda within our dialogic model.

Lois Gibbs's two-year struggle at Love Canal is well documented elsewhere.[57]

Here, we emphasize CCHW's evolving self-perception, as revealed in the language of its newsletters and other publications.[58] Again, the choice of words is important not only for understanding the communitarian discourse conveyed in the following quotes, but also for what it reveals about Gibbs's struggle to expand that vocabulary. The starting point of the group's analysis of the hazardous waste problem was that neither industry, government, nor even conventional environmental groups could come to grips with the problem:

> Toward the end of the 70s, the Environmental Movement was in trouble and its viability as a social movement was in doubt. Its main arenas were the courts, the corridors of state legislatures, Congress and the hundreds of bureaucratic cubbyholes where environmentalists tried to influence government enforcement agencies.[59]

Note the rejection of politics as usual in the above statement, most pointedly the repudiation of alleged public interest lobbying as a political strategy and even a questioning of the efficacy of the entire environmental movement.

A second feature of CCHW's attitude was self-reliance and a commitment to commonsense local activism as a substitute for the failure of lobbying and litigation strategies. Here is the democratic wish's yearning for citizen self-government:

> Lois decided that if there was a solution, she and her neighbors had to find it for themselves. No outside "savior" was going to come in and solve this problem for them. [The Love Canal Homeowners Association] quickly set goals, developed strategies and began to question government investigating teams. They discovered that if they asked scientists simple, common sense questions and demanded answers in plain English, the community could understand what was happening. . . . They learned that it's o.k. to demand that public officials express themselves clearly and in plain language. These lessons changed the whole tone of the issue and gave people back their confidence in their own common sense, as well as a greater feeling of control over the situation.[60]

This statement reveals the grassroots toxics movement's specific attempt to overcome the communitarian language's weakness regarding the complex scientific and technical issues of modern life. It also reveals an affinity for the educational strategy of the right-to-know movement.

Lois Gibbs was overwhelmed with responses to her victory at Love Canal. In effect, CCHW drew its initial operating identity from Gibbs's reaction to her new fame. The following statement marks the passage of CCHW into Geiser's second stage:

People still called or wrote, often after making dozens of calls to upstate New York trying to trace her. Most people who contacted her confirmed what she already knew: the conventional way of solving problems doesn't work on toxics issues and there still weren't any national groups around who understood what people were going through and knew how to help fight to win. The . . . CCHW . . . was born out of those desperate calls and out of Lois's determination to give other people the kind of help she had gotten when she first started at Love Canal. In 1981 and for the next two years, CCHW was a room in her basement, a phone and files and boxes filled with information.[61]

Slowly, CCHW began to emerge as a networking operation, struggling toward Geiser's third stage, linking local groups and sharing practical advice with groups protesting toxics issues in their communities:

> We began to see exactly what it took to win (people and action) and what comprised a *losing* formula (relying on lawyers and technical experts). However, the "Movement" was really only a wide scattering of uncon- nected "hot spots." CCHW didn't have the resources or people to do much more than try to stay within a week of so of being current with the daily mail and call-back telephone messages.[62]

After roughly two years of operating reactively, CCHW began to formulate positions on toxics policy. This maturing of the organization came about in re- sponse to several factors. First, the Reagan administration's environmental poli- cies kept the hazardous waste issue in the limelight (and CCHW's activities along with it). Second, increasing technical, practical, and fund raising sophistication gave the organization credibility, clout, and long-term stability. Third, CCHW increasingly realized that the consequences of hazardous waste policy were ineq- uitably distributed. Their concern predated much of the recent scholarly interest in environmental racism,[63] largely because from the beginning CCHW had its strongest ties with lower- and middle-income people.

The following statement reflects Gibbs's acknowledgment of the limitations of the communitarian language and her understanding of the strategy suggested by the right-to-know movement. Its message stands in striking contrast to the usual connotations of the NIMBY label pinned on the movement by its opponents:

> As the Movement matures, we've seen it move away from the NIMBY syn- drome. Though most groups start out saying, "Not here, no way, take it somewhere else!" Over time groups shift to "NIABY"—"Not in Anybody's Back Yard!" Group leaders take advantage of opportunities CCHW provides to meet folks from "somewhere else" and begin to decide that they won't

let their solution become someone else's problem. This new dominant attitude thrives on the increasing, new information on alternatives.[64]

Pursuing this core strategy, CCHW became a key supplier of information for
community toxics protesters around the nation. By 1983, it had begun publishing
its quarterly newsletter, *Everybody's Backyard*, regularly and offering technical and
organizational manuals (self-help guidebooks), along with leadership development
conferences and on-site organizing visits, to the various groups it served. In 1984,
CCHW maintained a network of some 300 groups. This number swelled to more
than 1,000 by 1986.[65] As of the end of 1991, CCHW had five field offices in addition
to its central headquarters in Arlington, Virginia, and it served approximately
7,500 grassroots toxics groups.[66] To place the size of this network of groups into
perspective, the previously cited Perlman study referred to the existence of at most
8,000 citizen action groups in total in the late 1970s.

Clearly, the Grassroots Movement for Environmental Justice, as CCHW often
calls its network, represents a significant and enduring mobilization of local citizens. It is involved in virtually every local fight against HWDF siting, and its success
in blocking facilities, forcing cleanups, and changing industrial practices has been
remarkable.[67] In explaining this success, CCHW repeatedly referred to three features of its organizational strategy: (1) helping local groups help themselves;
(2) adhering consistently to themes of waste reduction, corporate and government
accountability, and a redefinition of health and safety on local terms; and
(3) exposing community groups to the national nature of their local concerns—
"from NIMBY to NIABY," as CCHW put it.[68]

CCHW's strategy during the decade covered by our study worked by relying on
variations of the three empowering themes of the right-to-know movement, mentioned earlier: access to information, a consistent front in opposition to industry,
and local orientation. These features gave the movement power because government and industry found it increasingly problematic to avoid making public revelations about their potentially hazardous practices. These revelations were then
networked nationally, so that it was hard for government and industry to seek out
soft spots (poorly organized or economically desperate communities) in which to
carry out their programs.[69] This made it more likely that government and industry
would reexamine their past practices. Essentially, by resisting one kind of redistributive policy—the siting of HWDFs, and particularly landfills in the early
1980s—the movement was able to force, from the ground up, consideration of
another kind of redistributive policy, the costs of which were borne by hazardous
waste generators through industry expenditure on new ways—for example, source
reduction—to deal with toxics.

In sum, CCHW's experience illustrates the mobilizing power of the communitarian vision of politics and policy-making. But the CCHW ideology transcends

the narrow politics of self-interest implied by the NIMBY label. More important, CCHW's evolution fits within a larger pattern of local mobilization around the issue of hazardous substances,[70] and this pattern consistently moves toward the establishment of a national agenda for dealing with toxics, thus expanding the horizons of the communitarian language beyond an exclusively local, nontechnical, and nostalgic focus.

This is the sort of participation demanded of citizens by the dialogic model we propose in this book. But before the prospects for such policy discourse can be entertained, we must consider the relevance of the final limitation of the communitarian language. How likely is it that the grassroots toxics movement will persevere as a democratic movement and achieve programmatic success? At one level is the formidable opposition of industry. At another level is the consistent thwarting of the democratic wish in American politics by the bureaucratic reality of American government. These two obstacles converge in the operation of federalism.

The Challenge of Federalism to the Toxics Movement

An essential irony of the grassroots movement's success in redirecting national hazardous substances policy from the ground up is the role that federalism has played in the implementation of that policy. Two factors make it possible for federalism to be manipulated from the top down to thwart the movement's goals. First, the overlapping centers of power in federalism may be exploited by the movement's industry opponents, and, second, the institutionalization necessary to maintain access to the multiple bureaucracies in these layers of government may negate the movement's vital immediacy at the grassroots. Both of these factors reduce the possibility of realizing the meaningful participation of local citizens necessary for a dialogic policy process.

OSHA's Hazard Communication: The Old Federalism

Perhaps the most striking example of the way federalism may be manipulated to thwart communitarian initiatives is the attempt by OSHA during the Reagan presidency to defeat local and state right-to-know laws. The right-to-know movement, as noted, began as an effort to convince OSHA to promulgate a chemical labeling standard. OSHA did not produce such a standard until the waning days of the Carter administration.[71] That standard was immediately withdrawn by the Reagan administration and eventually replaced in 1983, at the height of state and local legislative activity in the area. OSHA, with the explicit encouragement of industrial interests, promulgated the Hazard Communication Standard, to go into effect in May 1986.[72]

On the surface, OSHA's actions might be interpreted as a sincere response to its legislative mandate to protect workers from occupational hazards. In fact, the standard was aimed at stopping community and state right-to-know initiatives in their tracks, largely through manipulation of the complex federal structure of occupational safety and health regulation in the United States.[73]

OSHA accomplished this manipulation of federalism in the following manner: first, it drafted the weakest possible disclosure provisions, largely granting management the power to determine which toxics to label; second, it permitted the broadest possible exclusion of trade secret substances from the labeling requirement; third, it limited enforcement of the standard to the manufacturing industry, in which similar labeling practices often already existed; and, finally, but most important, it pursued the complete preemption of all state and local right-to-know legislation from the field of enforcement, meaning not only state and local regulation of manufacturing industries, but also other industries untouched by the standard, and even state and local regulations involving industry disclosure to the community and its public safety officials.

The intended effect of the standard was to codify the status quo ante of workplace hazard disclosure[74] and then, through an extremely broad interpretation of the Occupational Safety and Health Act (1970) preemption provision, to remove state and local right-to-know regulation from the field. OSHA's standard immediately became the subject of litigation,[75] which, for the most part, foiled OSHA's attempt at preempting community right-to-know laws and forced OSHA to rewrite the standard to cover more workplaces.[76] The challenges to OSHA's Hazard Communication and the expanded coverage included in later revisions were aided by the industrial accident in Bhopal, India, in December 1984.[77] Once again, a catastrophic, widely reported event had redrawn the boundaries of the new social regulation.

The Bhopal tragedy affected toxics regulation in America in other, more complicated ways as well, and we discuss these later. But the more relevant point here is that OSHA's ill-fated attempt to subvert local right-to-know initiatives demonstrates, once again, the curious role that federalism plays in the policy process. Once seen (in the 1970s) as a national guarantor of worker safety and health against the inconsistent performance of state and local workplace regulation, OSHA now became the opposite, an irony made all the more glaring because it occurred under an administration ostensibly committed to devolving regulatory authority to states and localities under the New Federalism.

The EPA versus North Carolina

The irony of the New Federalism reappeared in the EPA's attempts to override state legislation that set HWDF siting limits more strictly than the federal government did. In 1985, GSX Chemical Services, Inc., had proposed and received an EPA clearance for operating a hazardous waste processing facility near Laurinburg, in Scotland County, North Carolina. The plan was to treat hazardous wastes and then dump them into the Lumber River, which supplied drinking water for the town of Lumberton, in neighboring Robeson County. As Philip Shabecoff reported in 1989, "The proposal drew protests from residents and others around the state. The area has a large proportion of blacks and Lumbee Indians, many of them poor. Opponents charged that the waste plant was earmarked for this region because the local community lacked the political power to resist the threat to drinking water."[78]

But Robeson County residents surprised GSX. They organized against the facility and convinced the North Carolina legislature in 1987 to set limits on river discharges far below those permitted by the EPA, thereby making it difficult for GSX to proceed profitably with its planned facility.

GSX and the Hazardous Waste Treatment Council (HWTC)—an industry lobbying group—immediately sued the EPA, asking it to withdraw North Carolina's RCRA authorization to regulate hazardous waste disposal. Lee Thomas, then the EPA's administrator, instigated proceedings to withdraw North Carolina's authorization but dropped the effort because it was statutorily untenable (RCRA explicitly supports the right of states to enact laws more restrictive than the EPA's hazardous waste regulations) and because of congressional opposition. In the wake of North Carolina's success in resisting the GSX site, other southeastern states began legislating restrictions against HWDFs.[79]

The hazardous waste disposal industry pressured Thomas's successor at the EPA, William Reilly, to reopen the case, which he did, amid charges that he was unfairly influenced by Dean Buntrock, head of Waste Management, Inc., and William Ruckelshaus, former EPA administrator and now chief executive officer of BFI. These embarrassing allegations led Reilly to shift the case from the EPA's Region IV administrator in Atlanta, Greer C. Tidwell (known to be critical of North Carolina's law), to the EPA's Region IX administrator in San Francisco, Daniel W. McGovern. On June 2, 1990, McGovern sided with North Carolina.[80] The decision eventually led GSX to abandon its siting efforts in the state, despite promises from its newly elected Republican governor, Terry Martin, that he would deliver an HWDF site during his term of office.[81]

The case is a reprise of all the arguments, couched in their incommensurable languages, covered in our examination of HWDF siting. David Case, legal counsel for the HWTC, argued in 1988 that Thomas's initial decision to drop the case was political. He said,

> He [Thomas] cracked under pressure from our opponents. He refused to deal with the real issue here and punted. . . . [North Carolina] is not deal-ing responsibly with its hazardous waste. . . . Without these facilities, there will be no means for adequately disposing of the nation's considerable waste stream.[82]

Commenting on the same decision, Velma Smith of the Environmental Policy Institute, which provided legal assistance in Robeson County's struggle, argued that

> the decision preserved a citizen's ability to go to local and state government and get some environmental protection. . . . [GSX's and HWTC's aim is] to build their plants where they choose, how they choose and to the specs they choose. They want the right to dictate to local communities what is safe.[83]

Meanwhile, EPA's reopened investigation, according to the agency's John Cannon, was aimed at determining

> whether the North Carolina law was intended to protect public health and the environment, or simply a response to political pressures from citizens who did not want a waste-treatment plant in their backyard.[84]

Case's assessment reflects the managerial urge to remove "politics" from decisions about the "real issue" of the hazardous waste disposal crisis and fails to engage Smith's claim, with its origins in the communitarian desire, for direct democracy and popular influence on the uses of private property. Neither confronts the pluralist worry about the evils of NIMBYism, raised by Cannon when he implied that North Carolina might not have been willing to take its "fair share" of hazardous waste disposal.

The Democratic Wish and the Prospects for Dialogue

The Robeson County case, like the Hazard Communication struggle, eventually favored the locality involved, but our point, once again, is to illustrate how the overlapping layers of government may be put to use in an effort to frustrate grassroots initiatives. The outcome in both instances was affected by events external to the struggles themselves—Bhopal in the Hazard Communication case and media coverage of influence peddling in the Robeson County one—which cut in favor of local interests. There are examples of this use of federalism that cut the other way (see chapter 4). But beyond this impediment to grassroots initiatives is an array of obstacles that reveal the communitarian language's third weakness, that is, its ability to sustain citizen-inspired agendas over the long haul amid the established interests and institutions of the American political economy.

One set of obstacles to the viability of the democratic wish is the obvious challenges offered by opponents who have greater financial and legal resources than grassroots groups typically have. One of the more well known of these is the Strategic Lawsuit against Public Participation, or SLAPP suit, leveled by opponents of grassroots groups in the hope of chilling local fervor by raising the specter of litigation against them. Most SLAPP suits are thrown out of court, but not before they force grassroots activists, characteristically short of the legal talent and front money needed for litigation, to defend themselves in court.[85]

A second obstacle is the corporate-sponsored grassroots group, which seeks to negate legitimate local activity with so-called citizen actions of its own. According to the CCHW, the Committee for a Constructive Tomorrow (CFACT) is just such a group. CFACT, a corporate-funded organization based in Washington, D.C., appeared at a Pennsylvania meeting of the CCHW and the local Concerned Citizens of Pequea Township. CFACT representatives distributed flyers at the meeting which alleged that CCHW was pursuing

> a radical political agenda that's anti-business, anti-people, and just plain liberal. . . . [The CCHW and other grassroots groups] not only terrorize folks from coast to coast about everything from incinerators and coal plants to pesticides and landfills, but some are also using such issues as labor relations, racism, and homelessness to push their radical agenda.[86]

Such heavy-handed maneuvers to hinder grassroots mobilization are likely to be less effective at that task than is the sheer complexity of American government. Such complexity rewards persistent organization, a national agenda, and a broad coalition of interests. These are precisely the weaknesses that typically plague grassroots groups. The grassroots movement against toxics had a national agenda and networked coalitions of local interests as well as a sophisticated national organizing mechanism, but CCHW still depended upon the commitment of its leadership to operate a decentralized organization. And the organization's sustenance was still drawn from the continuing efforts of local activists spontaneously organizing citizens around well-publicized toxic threats in their home communities.

Our dialogic model contains prescriptions for legitimate social regulation that depend upon specific features of the communitarian language of politics used by the grassroots movement we have described in this chapter. Most important among these is a central focus upon active participation in a political community. Without "strong" citizens, the kind of broad-ranging and ongoing dialogue we imagine for determining social regulation can never occur. The key concern for our dialogic model of social regulation is, thus, the staying power of grassroots mobilization around toxics issues. But as James Morone suggests, "Reformist energy that might have been mobilized on more radical purposes [is] spent on (newly legitimated) conflicts within a narrow institutional context. . . . Potentially radical forces gain participation and focus on the limited (and limiting) conflicts of organizational life."[87]

One example, among many, of the "limiting conflicts" facing the grassroots toxics movement is something that on the surface would seem to be a culminating victory, the federal Emergency Planning and Community Right-to-Know Act, commonly known as Title III[88] (of SARA [1986]).[89] Under Title III, State Emergency Response Commissions, chosen by governors, are to designate local planning districts and Local Emergency Planning Committees (LEPCs).[90] LEPC membership must be drawn from five statutorily specified local groups: elected officials, public health and safety officials, media representatives, community groups, and facility operators and owners. The primary role of LEPCs is to gather information on hazardous substances in their communities and to plan for emergencies in the case of accidents involving those substances. In essence, then, LEPCs have become a crucial conduit of the information so vital for citizen empowerment in the local struggle over toxics issues.

On the surface, the design of LEPCs may look like the perfect embodiment of dialogic principles. In fact, it has one serious shortcoming. Congress provided no funding for LEPCs, neither for information gathering nor for technical analysis. As a result, the sole source of expertise for understanding most toxics and the risks associated with them is the industry representatives on the committees. Domination of LEPC expertise by the one interest with the greatest incentive to distort the truth in its favor is a violation of every principle of free and equal dialogue underlying our dialogic model of policy-making.

The structure of LEPCs creates a threat to continuing mobilization of citizens around toxic issues because opponents of such mobilization can now point to LEPCs as a sufficient substitute for independent citizen action. If, in fact, LEPCs are dominated by industry, their broadly representative composition may be used to pit a "consensual" LEPC understanding of toxics policy against the "radical" one advocated by the local grassroots group, thus discrediting the latter. It would be ironic indeed, but also perfectly consistent with the tragic side of the democratic wish, if a law so beholden to the grassroots movement for its content and passage were ultimately responsible for its demise.

The relevance of federal legislation to the prospects for meaningful policy dialogue about hazardous waste regulation offers a fitting conclusion to this chapter. The federal legislation regulating hazardous wastes has certainly been part of the problem. But it can be part of the solution, too, if legislators begin to recognize the structural and procedural changes in the policy process that are necessary to promote regulatory dialogue. Hadden puts the point succinctly:

> Government may have to help citizens interpret or manipulate the data they obtain in order to make it germane to community decisions, not just to ensure its availability. The larger role of government does not necessarily lead to a reduced role for citizens; on the contrary, because the purpose is to allow citizens to make better decisions, there is a great burden on citi-

zens to evaluate information and actually make choices rather than leaving them to government or industry. . . . Government must also create or support institutions for participation and decision making that will help citizens make use of their new information.[91]

We have examined the evolution of the grassroots movement against toxics from a purely local phenomenon to a national attempt to reorient federal hazardous substances policy. But this refinement of the democratic wish can be only one voice in a dialogue made up of very different interests speaking in different languages and with varying degrees of influence within the policy process. Under these conditions, a free, equal, and knowledgeable exchange of views requires a recognition of and compensation for the structural differences and organizational vulnerabilities of the participants.

In particular, the communitarian voice remains most vulnerable to the rigors of sustained expression in a national dialogue over toxics policy. In contrast, the more conventional managerial and pluralist languages find easy expression amid the structure of federalism because they are the voices government and industry are accustomed to using and citizens are most used to hearing. Unless something challenges the dominance of these two languages in the policy dialogue, the grassroots toxics movement will become nothing more than another democratic wish smothered by a hostile bureaucracy, and it will be further than ever away from realization.

Our concluding chapter treats the differing levels at which change must occur for true policy dialogue to begin. Although there are sizable obstacles to that development, we argue that reinventing government, redefining private property, and recasting the notion of citizenship can at least establish criteria for redrawing the regulatory boundaries among the state, democracy, and capital. Accomplishing any of these tasks may be an impossible legislative challenge, but the grim reality of the alternative—policy gridlock over hazardous wastes—gives us some license to speculate about redirecting the future of social regulation.

8 / Citizenship in the New Regulatory State

A decade's experience with hazardous waste policy charts a regulatory frontier filled with conflict over the appropriate boundaries separating market, state, and democracy. The conflict is expressed in discordant voices, none speaking a language that can fully accommodate the redistributive consequences of the new social regulation. These voices talk past one another because their different vocabularies interpret the ingredients of the new social regulation—its controversial nature, its redistributive consequences, and its implementation within a federal system—in fundamentally different ways.

The managerial, pluralist, and communitarian discourses are used by policymakers, interest groups, and local citizens in ways that mutually exclude other language's core values. Because of this, they all envision a different map of this new regulatory frontier, each positioning science, private property, and public power in controversial ways. Regulation in America has a history of accommodating managerial and pluralist debate over the proper positioning of these elements, but the addition of communitarian objections—from citizens mobilized by the redistributive consequences of the new social regulation—destroys the governing balance between the conventional languages of regulation.

At issue, then, is how to map the frontier so as to capture a more complete regulatory landscape for the future. Because each language offers only a limited vision of what is at stake in the new social regulation, that regulatory landscape—the legitimate positioning of market, state, and democracy—can be agreed upon only tentatively as a matter of ongoing dialogue among disputants. We do not believe that *any* single language, whether one of the three we have analyzed or some metalanguage that synthesizes elements of managerial, pluralist, and communitarian discourses, will emerge to settle the conflicts over social regulation. Instead, competing languages must be held in continuous communication with one another, so that they complement rather than exclude the others. But, given the contradictory structural location of each party to the debate, the prospects for actually engaging in a dialogic process of policy-making are not at all obvious or self-activating.

We devote our conclusion to speculating about the structural and attitudinal changes that might be required in order to establish a proper forum for our dialogic model. As a caveat before beginning that speculation, note that we have used the dialogic model as a standard of criticism to guide our investigation of what went wrong with hazardous waste regulation in the 1980s. Its usefulness should be judged by that and by the constructive debate that our suggestions in this chapter create for considering the future of social regulation. The dialogic model should not be measured solely by its potential for practical realization; nor is it designed to produce some sort of final consensus about social regulation. Our suggestions here derive from lessons learned in our theoretical discussion and in our experience in the field of hazardous waste regulation.

Establishing a Context for Policy Dialogue

We organize our speculations about the future of social regulation around the ongoing debate among environmentalists and democratic theorists over the appropriate role for a democratic government in environmental policy-making. Some commentators favor a centralist approach as the only solution, with a strong national government forcing compliance with environmental mandates from the top down (see chapter 4). Decentralists, on the other hand, prefer environmental initiatives to well up from the populace, rejuvenated by a new environmental awareness that would flow innately from the establishment of a truly democratic community.[1]

Rather than regarding these as mutually exclusive, we believe that strong central governance has a role in the new social regulation, and so does democratic renewal. The key is to place government in pursuit of renewal. But this means more than superficial legislative or executive mandates favoring the environment or citizen participation in government. It means fundamentally rearranging the structural relations among market, state, and democracy so as to protect the environment and, at the same time, nurture democratic community. To accomplish this, a strong national government must be enlisted to restructure the privileged place of private property in public policy deliberations, and a decentralist populace must establish new structures of democratic participation upon which a sense of community might be built, so that citizens might responsibly address complex issues transcending their own selfish interests.

Market and Government: Property and Power

One of the problems facing all forms of regulation in America is that they challenge at some level the operation of private markets by correcting market failures. But the market imprisons regulation by inflicting economic pain in the short run for policies created to produce long-term gain. As we have argued, this is doubly true of the new social regulation, which balances concentrated costs focused upon well-organized interests against disputed benefits to be enjoyed by future generations. The government's sensitivity to market pressure is reinforced by constitutional presumptions that protect private property from public control. Thus, the structural relation between market and government and the nature of regulatory politics converge in a preference for the relatively unrestricted use of private property.

Simply put, this preference tolerates the introduction of risks into communities without comment. When those risks are so concentrated that they become obvious to the targeted communities, citizens comment anyway, questioning the government's uncritical acceptance of industry risks absorbed by the community. Popular suspicion of industry and government solutions to the hazardous waste problem in America is roundly criticized as an unwillingness on the part of communities to accept their fair share of the risks of modern industrial society. Lost amid the charges of NIMBYism are criticisms of the industrial practices that introduced those risks in the first place.

Another way of looking at it is that communities are protesting the way government subsidizes those industries generating hazardous wastes through expanded disposal capacity. Expanding HWDFs makes disposal cheaper but also transfers the risks of disposal to the host community. Until grassroots protests became successful at limiting disposal capacity, neither industry nor government had seriously considered alternatives—like banning hazardous production processes—that might impose redistributive burdens upon industry. In fact, concerted opposition to right-to-know initiatives placed government on the side of industry in asserting that communities should not be permitted even to know about risky private enterprises, much less entertain the idea of banning them.[2]

The new social regulation places government in an uncomfortable position between too much democracy and too much capitalism. Our point is that traditionally and constitutionally, government has sided too often with capital. As long as regulation seeks solutions for industry problems by redistributing their risks to specific communities, these policies will be challenged precisely because of that preference for the well-being of industry over the well-being of the community. The problem, even for those who support a government predisposition in favor of industry, is that these challenges, rooted as they are in enduring themes in American political history, have proved especially effective at stopping government and industry in their tracks. These policies are thus rejected as illegitimate by the citizens who are forced to bear their burden, and policy gridlock prevails.

From our perspective, the first step toward a solution to the conundrum of legitimate regulation is for government to abandon its risk-tolerant presumptions protecting industry activity and to adopt a preference for risk-averse industrial operations, forcing industry to shoulder the burden of proving that its processes are safe enough for community consumption.[3] Shifting the burden of risk from community to private industry is a prerequisite for establishing a context of trust within which policy dialogue might occur. Given our understanding of the politics of federalism, however, state and local governments could never accomplish this shift. Not only would the affected industries challenge it in federal courts on constitutional grounds, but they would also be in a position to threaten an "exit" from jurisdictions pondering such a change.

A shift as fundamental as this would have to take place at the federal level, where nationwide consistency in enforcement makes the threat of industrial exit less persuasive.[4] The centralist argument for environmental protection is appropriate here because it invokes the strength and pervasiveness of a national government to change industry's environmentally destructive habits. Further, a constitutional argument for preferring community health over industry profits would probably prove more persuasive in federal court were it based upon congressional legislation or an argument from the solicitor general.[5]

There is both legislative and judicial precedent for such a shift. The National Environmental Policy Act of 1969 was intended to accomplish just such a shift in the federal government's priorities for considering government actions affecting the environment. And the U.S. Supreme Court has vindicated the 1970 Occupational Safety and Health Act's preference for worker health over the costs to private industry in the consideration of workplace regulations.[6] Granted, neither piece of legislation has fared well in practice because of executive and judicial hostility toward implementing their mandates during most of the last two decades. But, absent that hostility, there is no reason to exclude shifting the burden of risk from the realm of the governmentally possible.

Subtle shifts in jurisprudence have made it possible for citizens to attack the imposition of risks by industry through litigation, but relying on "toxic torts" to rebalance risks has many drawbacks. This sort of litigation is haphazard and fragmented. Winning or losing depends as much upon the vagaries of jury selection and trial tactics as on the merits of the case or the seriousness of risk being litigated. Because of the adversarial nature of the proceedings, media coverage is often inflammatory and inaccurate. But, when important information about dangerous industrial practices might be revealed, the industries involved can often seal the record as part of their settlement, so that information about risks is actually suppressed. The parallel development of medical malpractice litigation suggests caution in advocating private suits as the means for reorienting federal risk regulation.[7]

Complementing this centralist task of the federal government would be federal programs supporting citizen education and access to information consistent with the government efforts called for by Susan Hadden.[8] We discussed federal programs directed at citizen education and participation (see chapter 4) like the specialized information clearinghouses—the so-called Backup Centers—of the Legal Services Corporations. Another possibility would be to restore and upgrade Congress's Science for Citizens program, which languished during the 1970s. Federal efforts like these would give citizens independent sources of information and the ability to analyze that information as it applied to the health and safety issues raised by industrial activities in their communities.

All of these programs would produce collective goods too expensive and too comprehensive for any state or locality to attempt. But even the most successful federal efforts to promote citizen command of information will be useless in the absence of citizens willing and able to take command of that information and inspired enough to use it for something more than their own interests. These are matters of socialization and attitude, things the federal government cannot supply from the top down; instead, they emerge from citizen activities at the grass-roots, and structural changes must occur there, too, in the nature of political participation.

Government and Democracy: Strong Citizens

Those who advocate a decentralist approach to reconciling environmental protection and democratic politics find a home at this level of our restructured landscape of social regulation. Their emphasis on renewed environmental citizenship can build directly upon the federal government's leadership in establishing a context of trust and citizen-friendly information in the regulatory process. Above we identified Benjamin Barber's *Strong Democracy* as a key argument in favor of our dialogic model (see chapter 3).[9] It is no coincidence that his suggestions for invigorating citizenship in America are relevant here in our conclusion. His reforms are indicative of the extent and holistic nature of the changes we feel are necessary at the grassroots level, if responsible citizen inclusion in an open dialogic process of policy-making is to occur.

As utopian as many of these reforms may have seemed in 1984, when Barber's book appeared, the events surrounding the presidential election of 1992 and the early days of the new Clinton administration indicate that diverse voices both inside and outside government have been confronting, albeit in a halting and incremental fashion, the problems of constructing a democratic citizenry capable of dealing with the challenges posed by rapid economic and technological change. During the campaign, both Clinton and Ross Perot called for the use of electronic town meetings as a way of actively engaging citizens in national political affairs.[10] The Clinton administration has called for and Congress has passed, although in a severely scaled down version, a National Service Act to allow college students to repay their school loans through public service. The stated reasoning behind this proposal was the need to foster a sense of public purpose among young people.

Similarly, as a way of fostering civic commitment, many high schools now require that students do volunteer work in their communities before they graduate.[11] More recently, in a highly publicized event, the Clinton administration unveiled its plans for "reinventing government," a set of proposals designed to restructure the federal bureaucracy. This plan was an explicit response to the belief of many in the new administration that government, especially the national government, was regarded by large numbers of the American people as incapable of solving the problems that confront the nation.

Many of these proposals may wind up having little impact, but they speak to the deep concern among national political elites over the political disengagement of many citizens and the resulting delegitimation of public policy initiatives. Until these problems can be addressed, the ability of government to confront public problems will be severely constrained. So, whereas reconstructing democratic citizenship may have once been the province of seemingly impractical political theorists, it is now a real problem confronting policymakers. From our perspective, however, the solutions articulated by national elites are lacking in two respects. First, they misread the scope of the task they have defined. Constructing a democratic citizenry adequate for the challenges of the next century is an enormous task, one that cannot be accomplished through piecemeal reforms. The great strength of Barber's democratic vision, and the reason we describe it in detail here, is that he outlines the sort of far-ranging and interconnected reforms that are necessary.

Second, the reforms proposed by national elites tend to ignore the enduring tradition of community-based political participation. A sensitivity to the vocabulary of the communitarian discourse, generally lacking among elites, is a prerequisite for understanding and drawing upon this tradition, and any successful reconstruction of citizenship must include and build upon this locally based democratic impulse. A strength of Barber's book is that it draws upon the communitarian language in its blueprint for democratic reform. We now turn to his specific plans for strong democracy.

1. *Neighborhood Assemblies*—The central element of strong democracy is a place where citizens can learn to engage in ongoing public dialogue about the issues confronting them. To fulfill this task, Barber advocates creating regularly scheduled public forums located within communities and modeled after New England town meetings. Trained professional facilitators would guide discussions aimed at involving all participants. Citizens would have the opportunity within these neighborhood assemblies to develop the communicative competence required for taking part in dialogic policy-making.

Barber's thought here is rooted in what we have been calling the dialogic model of politically relevant truth. Central to the dialogic model is the enormous difference between the opinions one forms in private contemplation, positions one need not defend with reference to any collective concerns, and the views one adopts when forced to defend and debate such issues in public forums.[12] Public debate and discussion is a requirement of citizenship because in such debate one is forced to acknowledge and address the positions of others rather than simply assert a rigid position. Considering seriously and in public the position of others is a critical step in fostering a dialogic sense of the public good. In contrast, the narrowly construed advocacy of individual private interests is encouraged when the ultimately passive processes of opinion polling and voting are the primary ways for citizens to articulate their preferences. This sort of opinion formation only emphasizes private concerns over common interests among citizens and assures a selfish vision of the public weal.[13]

2. *Two-way Communications Technologies*—In order to address issues affecting constituencies larger than the community, neighborhood assemblies would be connected to larger forums through the use of interactive electronic communications. As if anticipating criticisms that have attacked Perot's proposals for electronic town halls, Barber emphasizes that such technologies must be employed only after citizens have gained experience in public dialogue at the local level. If such experience is not provided, mass media deliberation may serve to mobilize groups around controversial issues and to block specific policy proposals, but they will remain unable to develop more constructive and enduring courses of action or habits of citizenship. Instead, such technologies will remain what they are now: either high-tech gimmicks with little political impact or a resource for a small number of already engaged elite political "junkies."

3. *Right of Access to Information*—Consistent with our discussion throughout this book on the importance of information to citizen empowerment, Barber's proposals would guarantee access to the information necessary for enlightened debate. Open access to computer information networks and other electronic data banks has the potential for fostering public dialogue and anticipates Clinton's campaign promise of "information superhighways."[14] Especially significant here is that Barber stresses the importance of scientific and technical information, often absent from unreformed communitarian rhetoric. Like Dewey, Barber includes science expertise as an integral part of democratic discourse. He demands only that it be made interpretable within the contours of a free, uncoerced dialogue among citizens, along the lines suggested by Susan Hadden in our conclusion to chapter 7.

4. *Greater Reliance on Initiative and Referendum*—As citizens acquire more experience in publicly discussing pressing issues, Barber would place greater reliance on initiative and referendum to decide important national issues. In this way, voting could be a direct extension of participatory democracy, rather than an indirect means of deciding public questions through elected representatives. Vital to the success of this method would be procedures for submitting issues in all their complexity to public vote by having both multiple options presented and multiple

stages in the voting process, rather than providing simple yes/no options on complicated one-shot referenda.[15]

5. *Universal Citizen Service and Choosing Officeholders by Lot*—Although Barber's first four proposals address the development of political structures and practices that provide opportunities for democratic participation, a second set describes mechanisms for increasing the number of individuals who will actually take advantage of those opportunities. These proposals indicate the degree to which creating public dialogue and an informed citizenry entail very fundamental changes in the way we think about political participation and representation. Given these requirements and the degree to which they have been ignored by policymakers, it is no wonder that actual participation, when it occurs, falls so far short of our expectations of an enlightened citizenry.

Creating citizens able and willing to take part in public dialogue is a prerequisite for a successful democratic political system. Achieving this requires civic education that emphasizes the obligations and rewards of participation in public life. This sort of education would provide the tools and motivations necessary if large numbers of individuals were to take advantage of the opportunities to participate in neighborhood associations, referenda, and large-scale public dialogue. Barber argues that mandating a period of universal public service would be an effective way to provide such a civic education to all citizens. There would be many options for the type of service an individual would perform, but everyone would be required to spend an extended period of time performing some type of public service. As we have seen, similar programs have been put forward by the Clinton administration and are now in place in many high schools. Barber's perspective reminds us that such proposals are only a small part of the larger task of constructing strong citizens.

A related proposal for increasing the number of individuals with experience at participating directly in public affairs is choosing local political officeholders by lot. A necessary component of any program aimed at creating greater direct participation in public issues is providing individuals with experience in managing public sector problems; selection of certain officials by lot would increase the pool of citizens having this experience and offer an incentive, since anyone might be chosen for office, to become familiar with the public issues confronting their communities.[16]

Such a proposal may seem farfetched, but it does dramatize the problems of existing representative institutions. These problems reflect what Morone calls the "yearning" in American political thought to bring government closer to the people: this yearning underlies both the democratic wish and the remarkably successful spread of term limitation initiatives in the states. The legitimacy of representative institutions is repeatedly challenged by local citizens who often see their own interests as being ignored or betrayed by their elected representatives (see chapter 6). That officeholding by lot or rotation is not practicable is certainly no vindication for existing means of representation.

6. *Workplace Democracy*—Consistent with our connection of workplace and residence (see chapters 4, 7), Barber expands the horizons of local political reform to include the workplace. Because people spend much of their lives at work and the well-being of their families and communities depends upon the circumstances of their employment, opportunities to share in making workplace decisions must be a central component of any program for increasing political participation.[17]

The speculations presented here are undeniably sweeping in their scope and in the levels of political life that they embrace. They focus primarily upon strengthening the communitarian aspects of public debate over social regulation, primarily because the communitarian voice has been consistently repressed or discounted within the current managerial-pluralist regime of regulatory policy-making. And its absence from policy dialogue is at least proximately responsible for the collapse of hazardous waste policy in America, as those left out of the process effectively veto its programs at the local level. By arguing for the inclusion of communitarian language in policy dialogue, we are by no means either discounting the importance of the managerial emphasis on scientific and technical expertise or ignoring the pluralist support for group competition and fair rules of the game. This should become apparent as we apply our speculations to the problem of HWDF siting.

A Dialogic Process for Siting HWDFS

The combination of centralist and decentralist reforms outlined above can be applied to inform our very real-world problem of policy gridlock surrounding HWDF siting. Indications from the few studies done of successful siting suggest that the dialogic model might provide an alternative framework for a rebuilding of the siting process. Before reviewing the studies that shed light on our model, we first enumerate the requirements of a siting process consistent with our discussion above.

First, and most important, citizens must have broad, ongoing opportunities to take part in public decision making about the generation and disposal of toxics. Limiting public participation to siting—especially when even last-minute participation must be forced on policymakers by grassroots NIMBY groups—is a guarantee that policy gridlock and failure will result. Instead, public officials must find ways of proactively bringing a broad range of citizens, not just the leaders of already existing interest groups, into an ongoing public discussion about the general benefits and costs of technological change as well as the more specific issues surrounding the generation and disposal of toxic substances. The failure of most citizens to join in such discussions cannot be simply laid at the doorstep of satisfied, apathetic, or ill-informed mass publics. Instead, the blame for this failure must be shared by political and economic elites as well as by the citizens themselves. Moreover, the absence of participation needs to be seen as a problem that must be overcome before successful policy-making can be expected. Industry's assumptions about its autonomy vis-à-vis hazardous practices must be suspended, and government preferences for the health and safety of the community must be asserted, if the dialogue is to be wide open. It is here that the national government must reinvent ways of giving states and localities the wherewithal—legal and financial—to overcome the structural limits imposed on them by federalism. Otherwise, establishing trust among citizens at the local level will be impossible.

Second, siting should be preceded by extensive public education about hazardous waste and the dangers of failing to regulate it safely, as was attempted, for example, by Florida's Water Quality Assurance Act and Amnesty Days. Public education *must* be considered a precursor to participation in decision making and not conceptualized, as in the managerial language, as a way of legitimizing the delegation of authority to experts or of gaining support for decisions that have already been made. Consistent with the decentralist concern for encouraging environmental consciousness among citizens, such an educational program would move citizens beyond their individual concerns about hazardous waste and confront them with their responsibilities to the larger community. Of course, this is a reciprocal process in which the larger community (as represented by government) would be equally obligated to demonstrate its commitments to the individual citizens.

Third, citizens and businesspeople (including disposal firms) must be brought into the process early on by credible public officials. This step should involve the removal of substantial structural barriers that prevent many affected interests from participating in policy-making. Here, our discussion of organizational and informational assistance for local groups comes to the fore, with the important addendum that inclusion of these interests must occur before any policy outcome is a foregone conclusion. Legitimation of toxics policy is impossible, we have found, if it is presented as a fait accompli.

Fourth, and relatedly, a wide range of alternatives must be taken into account, including alternatives that burden all parties in different ways—not just land-based disposal, for example, but also source reduction. This is, of course, consistent with several points made earlier but deserves attention here because it stresses how wide-open discussions can combine with a governmental preference for risk-averse technological alternatives. So often in siting debates, industry's "sunk costs" in one technology are used to exclude consideration of other, more responsible technologies. This sunk-cost argument appeared recently in the test-burning controversy surrounding the hazardous waste incineration plant in East Liverpool, Ohio.

Fifth, citizens must be guaranteed substantive, meaningful input into the study of potential sites, the siting itself, and the design and management of the site, once it is selected. This reform relaxes private industry's traditional control of its investment decisions and forces those decisions, as they introduce risks to the community, to be thought of as privileges, not rights. As privileges, they would be conditioned by local involvement. This means that the line between public and private would dissolve in every phase of the operation.

Sixth, and finally, the site operator, whether public or private, must be made to share in the fate of the communities affected by the site. This includes extensive liability provisions protecting the community and significant tax revenues funneled into the community from the operation of the site. This and the preceding requirement are aimed at assuring a "community of interest" between the facility and the locality, a point reinforced by our discussion of Elliott's research (see chapter 6).

No siting process, as far as we know, has ever met all of these conditions; but siting processes that have been successful did incorporate some of them.[18] Barry Rabe describes how facility siting in Alberta, Canada, was successful in 1987 after the province and private interests pursued a parallel strategy of site selection, combining both market-driven involvement of private disposal firms and province-driven selection of sites.[19] The province involved the public early on in preparing a list of suitable sites. The process entailed many public meetings and considerable education of the citizens involved. Following this, the province invited communities to compete for the site. Five communities were eager to do so. Once one was selected, a shared system of management was devised to give government, residents, and private firms a role in the ongoing operation of the facility. The facility was built and continues to operate in a remarkably cooperative way. As evidence that the process in Alberta was not a fluke, Canada managed to site another facility in 1992 through the same process.

Aspects of hazardous waste policy in California approximate some of the conditions of the model we have outlined.[20] After years of policy gridlock, progress has been made among the various factions in the dispute over hazardous waste disposal there. The complications in that state include an enormous hazardous waste problem, layers of government at first unwilling to cooperate, powerful industry forces, and an extremely sophisticated and active environmental movement. After years of going nowhere, the various factions formed a series of interlocking groups that led to dialogue about the prospects for solving the problem. The long-term association of members in these groups and the ultimate intractability of the hazardous waste problem, according to some observers, forced participants to go beyond self-interest and consider options that resulted in breaking the gridlock.[21]

Both of these cases have their limitations. For example, Minnesota faithfully employed the Alberta siting strategy and got nowhere. Indeed, Rabe remains pessimistic about prospects for following the Canadian model in the United States. Attempts to adopt procedures similar to those in Canada fail because of differences between the federal structures in the two countries. Significantly for our argument, the national government in the United States is unable or unwilling to "protect" state and local decision processes by barring the shipment of hazardous wastes from states that have not sited a facility to those that have. Thus, states are punished for acting responsibly. In contrast, Canadian provinces are able to restrict wastes at facilities to those generated within that province. This difference emphasizes the point we have been making that, in a federal system there is an intimate connection between the actions of a strong national government and the abilities of localities to make fair and open democratic siting decisions.[22] Further, California has yet to produce anything approximating a substantive outcome like a siting decision. Research by Portney is fairly pessimistic about the prospects of any policy mechanisms producing results that would avoid rejection on the part of locals.[23] His experience studying the relatively inclusive process implemented in Massachusetts (which has also failed to site an HWDF) and surveying residents there reveals a basic reluctance to permit siting under any circumstances.

But the failures of existing siting processes, no matter how inclusive, must be understood as being rooted within the context of the current landscape of social regulation. For progress to be made toward a responsible hazardous waste policy, the structural relations of that landscape must be changed before the behavior of its components can be effectively evaluated. Until we have a new sense of political community in which industry, government, and citizenship are redefined, we will not be able even to consider, much less endure, the severe policy choices presented by the new social regulation.

Appendix / Empirical Analysis: State Environmental Regulation in 1980

The following is a more detailed and technical description of the fifty-state study discussed generally in chapter 5. For public expenditures in the states on environmental quality control, we created our first dependent variable, PUBLIC, a straightforward measure of state-funded land and water regulation during fiscal year 1980.[1] For private expenditures during the same period, we recorded private spending on land and water pollution abatement, controlling for the size of polluting industries within a state.[2] This became our second dependent variable (PRIVATE), calculated by dividing the total dollar amount spent on land and water pollution abatement by the total value-added of the seven industries that produce the greatest amount of hazardous waste.[3]

Propositions and Data

Below are detailed descriptions of each of the independent variables introduced in our discussion in chapter 5. As we introduce them, we explain their theoretical significance with reference to the regulatory delegation theme of the New Federalism, on the one hand, and to our discussion in chapter 4 of the redistributive limits of a federal system, on the other.

1. *Need*. One of the tenets of regulatory delegation is that states can respond more flexibly to the relative severity of their social problems than can the federal government. Thus, states with relatively severe hazardous waste problems, in response to public pressure, should be in a position to impose stricter environmental regulations upon industry, translated into greater public and private expenditures on pollution. Our arguments in part I suggest that in spite of the seriousness of any social regulatory problem, states and localities will resist such pressure to burden industry or will attempt to transform its solution into an allocational issue drawing on general revenues; thus expenditures would depend more upon the fiscal capacity of the state than upon the size of the problem, as discussed below. Our conclusion is that the objective need for regulation will be associated with

neither public nor private expenditures. To measure need, we created a variable (NEED) based upon an EPA survey of hazardous waste sites in all fifty states, counting all unmonitored sites in each state as well as all unlined sites within one mile of a source of drinking water.[4]

Proposition 1: NEED will not be a determinant of PUBLIC or PRIVATE.

2. *Fiscal Capacity.* As the variable (STATEXP) indicating the total public resources available to state government, we use the size of states' total budgets for fiscal year 1980.[5] As discussed in chapter 4, there is great pressure on state and local policymakers to transform redistributive issues into allocational ones to avoid burdening private interests. Our approach suggests that, when pressures for regulation arise, state and local policymakers will avoid policies that impose redistributive costs on private industries. Instead, they will tend to pursue policies financed by state government, thereby translating demands for redistribution into allocational claims on the public treasury. Thus, the size of a state's environmental expenditures should vary directly with the size of the state's general revenue fund, that is, its fiscal capacity. By comparison, private industry expenditures should be unrelated to the size of the state's fiscal capacity.

Proposition 2: STATEXP will be a strong positive determinant of PUBLIC.

3. *Quality of Hazardous Waste Laws.* The argument for delegating regulatory power to the states might have pointed to the state laws regulating hazardous waste that existed in 1980 as an indication of the states' readiness to assume redistributive responsibilities from the EPA. This argument would suggest that the stronger the laws, the greater the public and private expenditures on the environment. Our discussion in chapter 4 counters by arguing that legislation which addresses redistributive issues is often symbolic and has little financial impact. Strong laws do not necessarily translate into budget allocations or enforcement pressure on private industry. As an indicator of the quality of a state's hazardous waste laws, we use the ranking (LAWRANK) of the inclusiveness of such laws compiled by the National Wildlife Association.[6] States are ranked relative to one another—the state with the most inclusive laws is scored *1*, the next most inclusive *2*, and so on.

Proposition 3: LAWRANK will be a determinant of neither PUBLIC nor PRIVATE.

4. *Industry Strength.* In chapter 4, we suggest that states will avoid redistributive policies for fear of discouraging economic development and the hoped-for expansion of state tax bases. Our argument suggests that the greater the economic significance of polluting industries, the greater the tendency for state policymakers to limit the redistributive impacts of regulation. In order to gauge the possible negative economic impact of industrial emigration—industry's ability to threaten the state with "exit"—due to hazardous waste regulation, we develop two indicators: first, the economic significance of industry (ECON) is calculated as the proportion of a state's total value-added by manufacture attributable to the seven industries that produce the most hazardous waste;[7] and, second, the importance of

these industries to the state's workforce (EMPLOY) is calculated as the total number of employees in these seven industries as a proportion of the state's total population.[8]

Proposition 4: ECON and EMPLOY will be negative determinants of both PUBLIC and PRIVATE.

5. *Industry Structure.* The discussion in chapter 4 and references from our interviews indicate that large and small firms respond differently to social regulation. Compared to small firms in highly competitive markets, large firms are able to consider a broader range of factors in their decisions: public relations, long-term investment in disposal equipment and facilities, and the possibility of future liability, among other things. Further, if the costs of regulation are high enough to discourage new entrants into the market, an industry composed of large, oligopolistic firms might actually support regulation as a means of reducing competition. As an indicator of industry structure, we measure the size of firms in polluting industries (CONCEN) as the proportion of businesses in each industry with twenty or more employees, averaged over the seven industries.[9]

Proposition 5: CONCEN will be a positive predictor of both PUBLIC and PRIVATE.

6. *Public Interest Group Strength.* Regulatory delegation implies that states will balance interests in the development of regulatory policy better than the federal agencies can, given the agencies' remoteness and undemocratic character. Significant environmental activism at the state level should, then, translate into demands for industry to take responsibility for pollution control. Our arguments suggest, on the contrary, that legislators will translate this activism into allocational rather than redistributive regulatory policy. Therefore, environmental activism in a state should lead to higher public, but not private, spending on environmental regulation. That is, industry spending on the environment should not be affected by the level of environmental activism. As an indicator of the strength of environmental activism mobilized around the issue of hazardous waste regulation (ENVGRP), we use the total statewide membership in the Sierra Club and Audubon Society as a proportion of a state's total population.[10]

Proposition 6: ENVGRP will be a positive determinant of PUBLIC, but unrelated to PRIVATE.

7. *Political Variables.* Our discussion in chapter 4 indicates that the ideology of policymakers should have little impact on their willingness to implement redistributive policies in the states, again calling into question the New Federalism's assertion that states are better able to balance competing political demands than federal agencies. If states were capable of making redistributive decisions, then partisan influence might produce two results. On the one hand, states controlled by liberal and regulation-oriented Democrats might be expected to support both public and private spending on environmental regulation. On the other hand, in states with competitive parties and uncertain election outcomes, politicians might

seek to placate public fears about pollution by spending more on regulation and passing laws that required the same from private industry. Our analysis indicates that these variables should be unrelated to expenditures on environmental regulation.

In order to examine this prediction, we include two measures of party control in states: Ranney's index of Democratic party strength (DEMO) calculated for the years 1974–80;[11] and a folded Ranney index of interparty competition (COMPET) for the same period.[12] DEMO takes on a value of 1.0 when there has been total Democratic success and a value of 0.0 for total Republican success. A score of 0.5 indicates a perfectly competitive two–party system. By subtracting all scores over 0.5 from 1.0, we transform the index into a measure of interparty competition: COMPET takes on a value of 0.5 for perfectly balanced parties and a value of 0.0 for total control by either party.[13]

Proposition 7: Neither DEMO nor COMPET will be a determinant of either PUBLIC or PRIVATE.

Data Analysis

To investigate the above propositions, we calculated a zero–order intercorrelation matrix among all variables and estimated two sets of regression equations:[14] the first set (equations 1 and 2) uses the level of public spending on hazardous waste regulation (PUBLIC) as the dependent variable; the second set (equations 3 and 4) uses private sector spending (PRIVATE) as the dependent variable. The independent variables used in all of the equations are ECON, EMPLOY, LAWRANK, ENVGRP, NEED, and CONCEN. COMPET is used with these variables in equations 1 and 3; DEMO is used with these variables in equations 2 and 4. The intercorrelation matrix appears in table 1;[15] regression results appear in table 2.

The adjusted R^2 for each of the equations in table 2 reveals that these variables explain a good deal of the variance in state expenditures for environmental regulation; but we are more interested in the patterns of relations than in the overall variance explained by the independent variables.[16]

Our analysis supports propositions 1, 2, and 3. LAWRANK's zero–order correlation with NEED (see table 1) indicates that states with pressing problems tend to respond by passing better laws; however, neither of these two variables is a significant predictor of either public or private spending on environmental regulation (see table 2). This finding is consistent with our expectation that politicians respond to pressures for redistribution by passing largely symbolic legislation.

The relations between NEED, STATEXP, and PUBLIC are particularly interesting. The zero–order correlation between NEED and PUBLIC is a highly significant 0.51 (see table 1), but when we controlled for STATEXP and calculated the partial correla-

Table 1. Intercorrelation Matrix

	Pub	Priv	Need	Empl	Envg	Econ	Conc	Law	Stat	Demo	Comp
Pub	1.00	−.26	.51	.63	.23	.35	.29	−.48	.82	.03	.26
	(***)	(.04)	(.00)	(.00)	(.06)	(.01)	(.02)	(.00)	(.00)	(.44)	(.05)
Priv	−.26	1.00	−.19	−.45	−.19	−.39	−.03	.12	−.25	.07	−.05
	(.04	(***)	(.09)	(.00)	(.10)	(.00)	(.41)	(.21)	(.04)	(.34)	(.37)
Need	.51	−.19	1.00	.42	−.13	.31	.16	−.39	.62	−.00	.14
	(.00)	(.09)	(***)	(.00)	(.19)	(.02)	(.13)	(.00)	(.00)	(.49)	(.21)
Empl	.63	−.45	.42	1.00	−.15	.58	.62	−.37	.55	−.00	.17
	(.00)	(.00)	(.00)	(***)	(.16)	(.00)	(.00)	(.00)	(.00)	(.50)	(.16)
Envg	.23	−.19	−.13	−.15	1.00	−.19	−.20	−.13	.04	−.21	.29
	(.06)	(.10)	(.19)	(.16)	(***)	(.10)	(.08)	(.18)	(.40)	(.10)	(.03)
Econ	.35	−.39	.31	.58	−.19	1.00	.57	−.41	.41	.27	−.13
	(.01)	(.00)	(.02)	(.00)	(.10)	(***)	(.00)	(.00)	(.00)	(.05)	(.22)
Conc	.29	−.03	.16	.63	−.21	.58	1.00	−.17	.22	.11	.09
	(.02)	(.41)	(.14)	(.00)	(.08)	(.00)	(***)	(.11)	(.06)	(.25)	(.30)
Law	−.48	.12	−.39	−.37	−.13	−.41	−.17	1.00	−.49	−.13	−.16
	(.00)	(.21)	(.00)	(.00)	(.18)	(.00)	(.11)	(***)	(.00)	(.21)	(.17)
Stat	.82	−.25	.62	.55	.04	.41	.22	−.49	1.00	.14	.24
	(.00)	(.04)	(.00)	(.00)	(.40)	(.00)	(.06)	(.00)	(***)	(.20)	(.07)
Demo	.03	.07	−.00	−.00	−.21	.27	.11	−.13	.14	1.00	−.47
	(.44)	(.34)	(.49)	(.50)	(.10)	(.05)	(.25)	(.21)	(.20)	(***)	(.00)
Comp	.26	.07	.14	.17	.29	−.13	.09	−.16	.24	−.47	1.00
	(.05)	(.37)	(.21)	(.16)	(.04)	(.22)	(.30)	(.17)	(.07)	(.00)	(***)

Significance levels in parentheses.

tion between NEED and PUBLIC, the result was a nonsignificant 0.01. The explanation for this lies in the fact that the zero-order correlation between STATEXP and PUBLIC is also strong (0.82, see table 1), as is the correlation between STATEXP and NEED (0.62, see table 1). What this means is that states with large budgets tend to spend more public money on environmental regulation than do states with small budgets, and that states with large budgets tend to have greater hazardous waste problems than states with small budgets have. But states with large budgets and great hazardous waste problems do not spend appreciably more on environmental regulation than do states that also have large budgets but relatively small hazardous waste problems. Thus, there appears to be an upper limit to public spending on environmental regulation, determined not by the size of the problem, as might be

Table 2. Regression Results

Independent Variables	Dependent Variables[a]			
	1 Public	2 Public	3 Private	4 Private
Statexp	+.69**	+.69**	+.02	−.03
	(50.92)	(50.31)	(.015)	(.019)
Need	+.03	+.03	−.04	−.02
	(.105)	(.093)	(.056)	(.016)
Employ	+.19**	+.17*	−.55**	−.52**
	(3.08)	(2.38)	(7.81)	(6.69)
Envgrp	+.40**	+.37**	+.07	−.09
	(12.32)	(11.62)	(.104)	(.193)
Econ	−.18**	−.16**	−.43**	−.45**
	(3.66)	(3.10)	(6.15)	(6.77)
Concen	+.32**	+.31**	+.52**	+.51**
	(7.08)	(7.23)	(5.64)	(5.74)
Lawrank	+.02	+.01	−.10	−.07
	(.051)	(.003)	(.378)	(.203)
Compet	.00	—	−.11	—
	(.000)		(.456)	
Demo	—	−.05	—	+.17
		(.306)		(.964)
R²	.82	.83	.40	.41
Adjusted R²	.79	.79	.28	.29

Note: We report the standard bets coefficients. F scores are in parentheses.
*Significant at .05 level
**Significant at .01 level
[a]Due to missing data, we were forced to exclude Nevada and Alaska from our analysis.

expected if states had the capacity to implement redistributive regulation, but rather by the size of the state's fiscal capacity, as predicted in our earlier discussion of the allocative tendencies of state regulatory financing.

Our analysis also confirms proposition 7, as the political variables have little impact on public or private expenditures on environmental regulation. Neither an ideological predisposition toward regulation (DEMO) nor uncertainty of electoral outcomes (COMPET) motivated state policymakers to regulate the environment. To guard against the possibility that conservative-Democratic, southern states were masking a relation between party control and regulation, we estimated equations without the eleven southern states, but our analysis revealed no relation between either partisan indicator and environmental expenditures.[17]

In contrast to the performance of the partisan measures in the regression equations, the level of environmental activism in the states, as measured by ENVGRP, is a significant predictor of public environmental expenditures (PUBLIC), although it has no effect on the amount private industry spent on pollution abatement. We take this result to be further confirmation of the theoretical discussion of redistributive capacity developed in chapter 4. Because states are unable to pursue redistribution, politicians (regardless of the partisan context) interpret environmentalists' demands in allocational terms. They therefore place the burden upon general revenue funds rather than taxes on industry, thus forcing environmental groups into competition with other interests for a share of the states' limited budgets.

The impact of the indicators of industry strength, ECON and EMPLOY, on public and private environmental expenditures partially confirms the prediction in proposition 4. Equations 1 and 2 in table 2 reveal that indeed the greater the significance of the affected industries to a state's economy, the less likely the state was to spend public funds or enforce pollution abatement costs upon the industries. Our emphasis on the primacy of economic development in regulatory decisions, discussed in chapter 4, is apparently borne out by the data. State policymakers are reluctant to impose redistributive regulatory burdens on industries important to the state's economy. The importance of labor in the affected industries (EMPLOY), which has a strong negative impact upon PRIVATE, as predicted in proposition 4, has an initially puzzling positive relation to PUBLIC. Perhaps this relation, which indicates that the larger labor's representation in a state's affected industries, the larger is the state's public expenditures on environmental regulation, can be explained consistently with the effect of environmental groups' influence upon environmental expenditures. Workers caught in the cross-pressure of environmental concern and economic self-preservation may have resolved this dilemma by encouraging policymakers to draw upon public funds for environmental regulation instead of saddling industry with costs that might threaten wages or even jobs. In effect, then, labor, like the environmentalists, influenced the tendency of state policymakers to use public funds rather than private enforcement to pay for environmental regulation.

Finally, in states in which the affected industries comprised large rather than small firms, both public and private environmental expenditures were greater, thus confirming proposition 5. When industries in a state were composed generally of large firms, they appeared more willing (and, of course, more able) to bear a redistributive burden for pollution abatement, but their size also affected the state's willingness to spend more on environmental quality control. When industries in a state were made up of generally small firms, neither the industries nor the states devoted resources to environmental regulation. This finding confirms the consistently held beliefs of environmental regulation officials whom we interviewed in Florida, Ohio, and New Jersey not long after these data were gathered. Small firms were considered, in some ways, a greater threat to the states' hazardous waste problems, despite the fact that they generated less hazardous waste than their larger counterparts in the industries regulated. Given their competitive positions and their sensitivity to marginal increases in regulatory costs, their incentive to dispose of hazardous waste improperly was greater than that of larger firms.

Notes

1 / Introduction

1. We discuss this law in more detail in chapter 6.

2. From our interview with the chairperson of the New Jersey HWAC (1983). For a full description, see chapter 6.

3. Quoted in Patrick G. Marshall, "Not In My Back Yard!" *Editorial Research Report* (Washington, D.C.: Congressional Quarterly, 1989), 314.

4. From our interview with members of the Ironbound Community (1983). For a full description, see chapter 6.

5. Our case studies are reported in detail in chapter 6.

6. "Economic regulation generally refers to the control of entry of individual firms into particular lines of business and the setting of prices that may be charged. In certain situations, it includes the specification of standards of service the firms can offer. Such regulatory measures are most justified when, because of the nature of the particular industry, only one or at most a few firms are capable of using their market power to engage in anticompetitive behavior." Robert E. Litan and William D. Nordhaus, *Reforming Federal Regulation* (New Haven: Yale University Press, 1983), 6.

7. See Martha Derthick and Paul J. Quirk, *The Politics of Deregulation* (Washington, D.C.: Brookings Institution, 1985).

8. David Vogel, "The 'New' Social Regulation in Historical and Comparative Perspective," in *Regulation in Perspective,* ed. Thomas K. McCraw (Cambridge: Harvard University Press, 1981), 238.

9. Litan and Nordhaus, *Reforming Regulation,* 10.

10. Arthur Okun, *Equality vs. Efficiency: The Big Tradeoff* (Washington, D.C.: Brookings Institution, 1976).

11. Robert E. Goodin flatly states that even attempting to reduce such values to dollar equivalents degrades them and lessens the possibility of achieving enlightened public policy. He illustrates the inability of price to capture fully such values when he notes that, no matter what the price established for contracting lung cancer, losing one's eyesight, etc., no one would freely choose to accept the money in exchange for the loss. Market transactions are simply unable to compensate individuals fully for such catastrophic losses. See his *Political Theory and Public Policy* (Chicago: University of Chicago Press, 1982). On this general topic, see also Mark Sagoff, "The Limits of Cost-Benefit Analysis," *Report from the Center for Philosophy and Public Policy* 9–11 (1982); Mark Sagoff, "We Have Met the Enemy and He Is Us or Conflict and Contradiction in Environmental Law," *Environmental Law* 12 (1982), 283–315. For a related argument about the inability of tort law to deal with such catastrophic losses, see Richard H. Gaskins, *Environmental Accidents* (Philadelphia: Temple University Press, 1989).

12. The reference to Thucydides is from Gary Wills, "Keeper of the Seal," *New York Review of Books* 38 (July 18, 1991), 20. Wills also provides the original quote from Thucydides: Cocyra "fell into its component parts" because "the agreed upon currency of words for things was subjected to random barter."

13. William E. Connolly, *The Terms of Political Discourse* (Princeton: Princeton University Press, 1983), 10.

14. Jean-François Lyotard, *The Postmodern Condition: A Report on Knowledge,* trans. G. Bennington and B. Massumi (Minneapolis: University of Minnesota Press, 1984), xxiii–xxiv.

15. Richard J. Bernstein, *The New Constellation: The Ethical-Political Horizons of Modernity / Postmodernity* (Cambridge: MIT Press, 1991).

16. Michel Foucault, *The Order of Things* (New York: Vintage Books, 1970), 44.

17. See, e.g., Michel Foucault, "Politics and the Study of Discourse," as well as the other essays in *The Foucault Effect: Studies in Governability,* ed. Graham Burchell, Colin Gordon, and Peter Miller (Chicago: University of Chicago Press, 1991). The quote summarizing Foucault is from Chris Weeden, *Feminist Practices and Poststructuralist Theory* (New York: Basil Blackwell, 1987), 35.

18. Even Jacques Derrida, the theorist most accused of treating everything as if it were simply a text, intends his analysis to be connected to political and institutional practices. Richard J. Bernstein, in *The New Constellation,* 211–12, makes this case: "The most common criticism of Derrida is that he 'reduces' everything to texts (and / or language) and declares that there is nothing beyond the text. But the key question is what does Derrida mean by a 'text'? . . . Derrida claims that his 'strategic reevaluation of the concept of text allows [him] to bring together in a more consistent fashion, in the most consistent fashion possible, theoretico-philosophical necessities with "practical" political and other necessities of what is called deconstruction.' He emphatically affirms that 'deconstructive practices are also and first of all political and institutional practices.," The Derrida quotes are from "But beyond . . . (Open letter to Anne McClintock and Rob Nixon)," *Critical Inquiry* 13 (Autumn 1986), 160.

19. John Fiske, *Television Culture* (London: Methuen, 1987), 14.

20. Indeed, in one of the better recent studies of the new social regulation, Richard A. Harris and Sidney M. Milkis recognize the limitations of their focus on national politics and acknowledge the need for an analysis of these issues at the state and local level. *The Politics of Regulatory Change* (New York: Oxford University Press, 1989), 302.

21. A much-publicized focus group study funded by the Kettering Foundation (1991) suggested that although many citizens are alienated from national politics and politicians, they make a clear distinction between politics at more distant levels and issues arising in their local communities, which they deem important and worthy of their attention and participation. See The Harwood Group, *Citizens and Politics: A View from Mainstreet* (Dayton, Ohio: Kettering Foundation, 1991). The issue may turn on the fact that citizens simply do not call what goes on in their community *politics* because the term has acquired such negative connotations. This suggests that we get a skewed idea of how citizens understand social regulatory politics if we focus only, as do most scholars, on the national level. On these points, see David Mathews, *Politics for the People: Finding a Responsible Public Voice* (Urbana: University of Illinois Press, 1994), esp. 1–5.

2 / The Search for the Public Interest in Three Languages

1. Charles A. Beard and William Beard, *The American Leviathan: The Republic in the Machine Age* (New York, 1931), 5, 8, 9, cited in Walter A. McDougall, *. . . the Heavens and the Earth* (New York: Basic Books, 1985), 72–73.

2. This language is based in what Robert R. Alford and Roger Friedland label the managerial home domain, which focuses on the implications of the growth of bureaucratic organizations for both the modern state and capitalism. From this perspective, the basis of legitimacy in the modern state is expertise rather than democratic consensus. In the benign, or "functional," version of the managerial perspective, expertise and the growth of bureaucracy are seen as creating the possibility of elite-managed rationalization of state and economy. The less benign version of the managerial perspective, however, emphasizes the power of large bureaucratic actors to extend control over social life and limit individual freedom. In either version, widespread participation and democracy become impossible owing to the fundamentally undemocratic nature of rational-

legal bureaucracy and the increasing complexity of modern society. See Robert R. Alford and Roger Friedland, *The Powers of Theory: Capitalism, the State, and Democracy* (Cambridge: Cambridge University Press, 1985), part II.

3. Such rules exert a powerful but usually ignored influence on regulatory policy. Originating in the study of organizational behavior, a growing literature in political science focuses on the importance of the culture or values embedded within political institutions as influences on current decision making. James March and Johan Olson, for example, call this approach the "new institutionalism" and argue that, in addition to the individual interests of current decision makers, the structure and history of political institutions exercise an important influence on behavior: see *Rediscovering Institutions: The Organizational Basis of Politics* (New York: Free Press, 1989). For a treatment of similar issues, see Steven Skowronek, *Building a New American State: The Expansion of National Administrative Capabilities, 1877–1920* (Cambridge: Cambridge University Press, 1982). Similarly, Eugene Lewis, in *American Politics in a Bureaucratic Age* (Cambridge, Mass.: Winthrop Publishers, 1977), calls the culture of policy-making institutions a "received value-mix" that guides decision making. Michael Cohen, James March, and Johan Olson (1972) laid some of the groundwork for this approach when they argued that rather than searching for solutions to current problems, organizations develop a set of solutions (derived from past experience) and then search for problems to which these solutions will be applied: see "A Garbage Can Model of Organizational Choice," *Administrative Science Quarterly* 17 (1972), 1–25.

4. See, for example, Alfred D. Chandler, *The Visible Hand* (Cambridge: Harvard University Press, 1977); Jerry Israel, ed., *Building the Organizational Society* (New York: Free Press, 1972); Jeffrey R. Lustig, *Corporate Liberalism* (Berkeley: University of California Press, 1982); Robert H. Wiebe, *The Search for Order, 1877–1920* (New York: Hill and Wang, 1967).

5. Richard Hofstadter, *The Age of Reform* (New York: Vintage Books, 1955), 225.

6. Russell L. Hanson, *The Democratic Imagination in America* (Princeton: Princeton University Press, 1985), chap. 7.

7. Ibid., 225.

8. Ibid., 245.

9. Indeed, Richard A. Harris and Sidney M. Milkis argue that the New Deal gave rise to "a new regulatory regime," quite distinct from the Progressive era's approach. Although we agree with Harris and Milkis that the New Deal represented a departure from the Progressive era, our purpose here is to demonstrate the continuity of the managerial language as one significant and continuing influence on policymakers' thoughts about regulatory policy. See their *The Politics of Regulatory Change: A Tale of Two Agencies* (New York: Oxford University Press, 1989).

10. Hanson, *Democratic Imagination*, chap. 8.

11. Likewise, democratic consumerism vests great power in private market actors. The fortunes of public officials depend on their ability to induce those behaviors from private market actors that lead to increased production and consumption. Thus, by firmly establishing the idea that government legitimacy was based upon fostering increased consumption, yet leaving production responsibility with private corporations operating in the market, New Deal political rhetoric in large part created the prison of the market described by Charles Lindblom in "The Market as Prison," *Journal of Politics* 44 (May 1982). In such a system, great power lies with those private market actors who can define the types of inducements they need to behave appropriately. It also makes government regulation of the market a problematic task indeed.

12. Thomas McCraw, *Prophets of Regulation: Charles Francis Adams, Louis D. Brandeis, James M. Landis, and Alfred E. Kahn* (Cambridge: Belknap Press of Harvard University Press, 1984), 212.

13. James Landis, *The Administrative Process* (New Haven: Yale University Press, 1938).

14. Ibid., 154–155.

15. McCraw, *Prophets of Regulation*, 302.

16. Hanson, *Democratic Imagination*, chap. 9. Viewed in this light, even programs like the Great Society may be seen as attempts to expand the benefits of consumption to previously neglected segments of the population. The limitations of the managerial perspective are exposed when such policies run into problems because those who are their beneficiaries at-

tempt to seize control from expert bureaucrats. See chapter 4 for a discussion of certain Great Society programs, especially CAPs, that ran counter to this managerial focus by attempting to empower poor people at the local level.

17. As we argue below, those who advocate "economic efficiency" as the criterion for evaluating regulation often begin by couching their arguments in the pluralist language. However, demonstrating the inadequacy of any single language for capturing the dynamics of social regulation, they wind up adopting the logic of the managerial perspective for the implementation of their policy initiatives. Equating the public interest with economic efficiency, they believe that rational standards imposed by trained experts (e.g., economists), rather than fair process, ought to be the hallmark of successful regulatory policy. Because so much of social regulation centers on issues that are scientific or technical in nature, granting policy-making authority to experts within bureaucratic state agencies is widely accepted. See, for example, Robert W. Crandall and Lester B. Lave, eds., *The Scientific Basis of Health and Safety Regulation* (Washington, D.C.: Brookings Institution, 1981); Lester B. Lave, *The Strategy of Social Regulation* (Washington, D.C.: Brookings Institution, 1981). Implicit in such approaches is the assumption that organizationally based expertise can provide the basis for the rationalization of social regulatory policy.

18. McCraw, *Prophets of Regulation*, 86.

19. McCraw, for example, notes that Brandeis's blanket abhorrence of "bigness" in the private sector led him to ignore differences between "center" and "peripheral" firms in terms of economies of scale, productive and allocational efficiency, and the possibilities and consequences of collusion. Center firms are in industries such as oil, steel, and automotive and are characterized by capital-intensity, advanced technology, mass production, and considerable economies of scale. Owing to the resulting high entry costs associated with such characteristics, oligopoly and natural monopoly are common. In contrast, McCraw in *Prophets of Regulation* notes that peripheral firms are "small, labor-intensive, managerially thin, and bereft of scale economies" (73). They exist in highly competitive markets in which collusion, oligopoly, and monopoly are extremely difficult to maintain.

While he may have been wrong from a strictly economic perspective, Brandeis raised points about the *political* power of large firms that are quite similar to the distinction between monopoly and competitive capital made by James O'Connor in his influential neo-Marxist treatment of American political economy, *The Fiscal Crisis of the State* (New York: St. Martin's Press, 1973).

20. Cited in McCraw, *Prophets of Regulation*, 108–9.

21. David Vogel, "The 'New' Social Regulation in Historical and Comparative Perspective," in *Regulation in Perspective*, ed. Thomas K. McCraw (Cambridge: Harvard University Press, 1981), 238.

22. Paul R. Portney, "Toxic Substance Policy and the Protection of Health," in *Current Issues in U.S. Environmental Policy*, ed. Paul R. Portney (Baltimore: Johns Hopkins University Press, 1978).

23. See, for example, Nicholas A. Ashford, *Crisis in the Workplace: Occupational Disease and Injury* (Cambridge: MIT Press, 1976); Daniel Berman, *Death on the Job: Occupational Health and Safety Struggles in the United States* (New York: Monthly Review Press, 1981).

24. Gio Batta Gori, "The Regulation of Carcinogenic Hazards," *Science* 208 (April 18, 1980), 256–61.

25. Edith Efron, *The Apocalyptics* (New York: Simon and Schuster, 1984); Samuel S. Epstein, *The Politics of Cancer* (San Francisco: Sierra Club Books, 1978); Sylvia Tesh and Bruce A. Williams, "Science, Identity Politics, and Environmental Racism," paper presented at the annual meeting of the American Political Science Association, New York, 1994.

26. Litan and Nordhaus, *Reforming Regulation*, 12–13; Gori, "Carcinogenic Hazards"; Lave, *Strategy of Social Regulation*, chap. 3.

27. Litan and Nordhaus, *Reforming Regulation*, 48.

28. Indeed, James Q. Wilson proposes a classification scheme for regulatory policies based on the nature of the groups adversely and positively affected. See his "Politics of Regulation," in *The Politics of Regulation*, ed. James Q. Wilson (New York: Basic Books, 1978).

29. See chapter 1.

30. Harris and Milkis, *Regulatory Change*, 244.

31. Cited in ibid., 266.

32. Alford and Friedland, *Powers of Theory*, 4. Our analysis of this language draws heavily upon their critique of pluralism.

33. Barry Karl notes, for example, the need of the Roosevelt administration to balance the political forces necessary to pass legislation (a pluralist concern) that would create the bureaucratic agencies central to Landis's kind of managerial vision. Indeed, legislative priorities often precluded managerial concerns, as in the failure to institute recommendations of the Brownlow Committee to centralize and rationalize executive authority. See Karl, *The Uneasy State* (Chicago: University of Chicago Press, 1983), 156–57.

34. For two of the most cogent and well-known elaborations of this position, see Milton Friedman, *Capitalism and Freedom* (Chicago: University of Chicago Press, 1962), and Friedrich A. Hayek, *The Road to Serfdom* (Chicago: University of Chicago Press, 1944).

35. Of course, once we question the origins of the preferences expressed in the market and admit that they can be systematically manipulated by powerful institutions, many of the advantages of the private market relative to the public sector disappear.

36. For example, if one owns land downwind from a factory, the costs imposed by emissions are not going to figure into the calculations of the factory's owner. Hence, the landowner will be forced to absorb these costs that result from an excess production of pollution. Clearly, this pollution violates property rights in the sense that it diminishes, without any compensation, the value of downwind land. This example poses a variety of problems. Conservatives tend to suggest that government action is warranted only when the costs of the pollution falling on all parties are greater than the amount that the factory owner would charge to reduce its production (i.e., if the externality is "pareto relevant"). Government action is warranted here because the benefits of regulation will exceed its costs, but markets fail because of the transaction costs involved in actually aggregating the costs that fall on individuals. However, there is a prior political issue, usually ignored by conservatives, having to do with the assignment of the burden of proof and the definition of property rights in such cases. For example,

do factory owners have the right to pollute up to the point of pareto relevancy, thus placing the burden of proving the level of costs on damaged parties; or must the factory owner prove that the benefits from production exceed the costs *before* polluting? That is, does existing usage of private property have priority over the costs imposed by changes in use patterns? The point here is that these are political questions, and the political decisions that answer them establish the context within which markets function (on the ways in which courts have changed their approach to the issue of prior usage, see Morton J. Horwitz, *The Transformation of American Law* [Cambridge: Harvard University Press, 1977]).

37. Milton and Rose Friedman make this point when they argue that various interests have turned away from competition in the private sector, which is socially efficient, to extracting benefits from government, which is not efficient. See their *Free to Choose: A Personal Statement* (New York: Harcourt Brace Jovanovich, 1980). Our notion that both conservatives and liberals employ the pluralist discourse is not so far from Theodore Lowi's argument about the establishment of interest group liberalism: see *The End of Liberalism* (New York: W. W. Norton, 1979) and our discussion below.

38. Mancur Olson, *The Logic of Collective Action* (Cambridge: Harvard University Press, 1965).

39. George Stigler, *Citizen and the State* (Chicago: University of Chicago Press, 1975).

40. Burton A. Weisbrod et al., *Public Interest Law: An Economic and Institutional Analysis* (Berkeley: University of California Press, 1977), 30–41.

41. Harris and Milkis, *Regulatory Change*, 290.

42. On this point, see Benjamin Barber, *Strong Democracy* (Berkeley: University of California Press, 1984); Robert N. Bellah et al., *Habits of the Heart* (Berkeley: University of California Press, 1985); and Olson, *Collective Action*. As both Barber and Bellah et al. make clear, such fears have deep roots in American political thought and go back at least as far as Tocqueville.

43. Indeed, such concerns have been a constant theme in debate over the possibilities and

meanings of American democracy since the founding. See, for example, the treatment of the thought of Orestes Brownson in Hanson, *The Democratic Imagination*.

44. Charles Lindblom, *Politics and Markets* (New York: Basic Books, 1977), and id., "The Market as Prison."

45. For examples of this type of approach, see Mark Green, Beverly C. Moore, Jr., and Bruce Wasserstein, *The Closed Enterprise System: Ralph Nader's Study Group on Antitrust Enforcement* (New York: Grossman Publishers, 1972); and Joan Claybrooke, *Retreat from Safety* (New York: Pantheon, 1973).

46. We find evidence of the negotiation process in the evolution of hazardous waste regulation described in chapter 6. Evidence of discrepancies in meanings occurred in the manipulation of terms and rhetoric by the Reagan administration to achieve its political goals (see chapter 5).

47. Shlomo Avineri and Avner De-Shalit, "Introduction," in *Communitarianism and Individualism*, ed. Shlomo Avineri and Avner De-Shalit (New York: Oxford University Press, 1992), 2–3.

48. Jennifer Nedelsky, *Private Property and the Limits of American Constitutionalism: The Madisonian Framework and Its Legacy* (Chicago: University of Chicago Press, 1990).

49. Ibid., 213–14 (footnotes omitted).

50. For an interesting treatment of the problems created in American political ideology by this conflict between the dynamic forces of capitalism and older "conservative" social values, see Gary Wills, *Reagan's America* (New York: Doubleday, 1987).

51. James Morone, *The Democratic Wish* (New York: Basic Books, 1990).

52. Brandeis's blanket condemnation of bigness, cited above, is an example of this suspicion.

53. An exception to this neglect is Barry Karl, who notes the degree to which the persistence of localism and the suspicion of a strong national government thwarted many of the goals of New Deal reformers. See his *The Uneasy State*.

54. There is a strong element of this line of reasoning in Brandeis's critique of bigness as a threat to democracy.

55. Indeed, from a Marxist perspective, the failure of reform efforts is explained by the fact that, while the communitarian language is as close as American political rhetoric gets to a systematic critique of American capitalism, it does not include a class-based analysis or prescription for economic reform.

56. Although Harris and Milkis, in *Regulatory Change*, argue that the new social regulation constituted a new regulatory regime that dominated policy-making until the Reagan administration, the major alteration they document represented an amalgam of managerial and pluralist reforms: the inclusion of public interest groups in the subgovernments influencing regulatory policies and attempts by Congress to guide more precisely the decisions of experts within regulatory agencies.

57. Theodore Lowi, *The End of Liberalism* (New York: W. W. Norton, 1979).

58. While critiquing the inadequacies of existing institutions, Lowi's proposals for reform, in our view, remain squarely within the pluralist discourse because he takes interest group struggle for granted as the central dynamic of politics and argues that what accounts for the crisis is the way this struggle has moved from the legislative realm—where it is appropriate—to the bureaucratic realm—where it is inappropriate. For Lowi, it is this shift itself that explains the pathology of public authority. Thus, his solution—"juridical democracy"—calls upon the courts to restore legitimate pluralist politics by guaranteeing that interest group struggle be restored to the legislative realm. Unexamined and outside this argument are the structural biases in that interest group struggle itself, or the communitarian call for greater direct participation by citizens.

59. Wilson, "Politics of Regulation."

60. Ibid., 370.

61. Litan and Nordhaus, *Reforming Regulation*, 93.

62. Ibid.

63. In general, as interest group liberalism suggests, social regulatory agencies are only vaguely controlled by other branches of government. After surveying attempts by Congress and the presidency to gain control of the regulatory process, Litan and Nordhaus conclude that congressional oversight is "generally weak

and sporadic," while executive oversight, even under the Reagan administration, was "more a goal than a reality." Ibid., 66, 81.

64. Harris and Milkis, *Regulatory Change*, 305.

3 / A Dialogic Model of Social Regulation

1. As we noted in chapter 2, the boundaries of postmodernism are extremely difficult to draw precisely. Many, for example, would not place Foucault within this category. This is why we chose the term *constellation*. Although Foucault may not be considered a postmodernist, he was clearly operating within the broader constellation of modernity / postmodernity. For an overall summary and critique of postmodernism, see Steven Best and Douglas Kellner, *Postmodern Theory: Critical Interrogations* (New York: Guilford Press, 1991). For a summary of postmodernism that highlights its relevance to social science, see Pauline Marie Rosenau, *Post-Modernism and the Social Sciences: Insights, Inroads, and Intrusions* (Princeton: Princeton University Press, 1992). On the relevance of postmodernism to political science, see Murray Edelman, *Constructing the Political Spectacle* (Chicago: University of Chicago Press, 1988). Among the works we relied upon most in coming to grips with this challenging literature were Jacques Derrida, "Structure, Sign, and Play in the Discourse of the Human Sciences," in *Writing and Difference*, trans. Alan Bass (Chicago: University of Chicago Press, 1978); Jean-François Lyotard, *The Postmodern Condition* (Minneapolis: University of Minnesota Press, 1984); Michel Foucault, *The Order of Things: An Archaeology of the Human Science* (New York: Vantage, 1970); Stanley Fish, *Is There a Text in the Class?: The Authority of Interpretive Communities* (Cambridge: Harvard University Press, 1980); Stanley Fish, *There's No Such Thing as Free Speech . . . And It's a Good Thing Too* (New York: Oxford University Press, 1994); Richard Rorty, *Objectivity, Relativism, and Truth* (New York: Cambridge University Press, 1991).

2. Richard J. Bernstein, *The New Constellation: The Ethical-Political Horizons of Modernity / Postmodernity* (Cambridge: MIT Press, 1991), 8–9.

3. A good example is the work of Jürgen Habermas, whom we discuss below. On the one hand, he explicitly challenges many of the arguments of postmodern scholars, especially Derrida. Yet, on the other hand, as Bernstein points out, Habermas's work is in constant dialogue with postmodernism and is difficult to understand if one does not first recognize the degree to which it operates within the constellation of modernity / postmodernity. For Habermas's critique of Derrida, see his *The Philosophical Discourse of Modernity*, trans. F. Lawrence (Cambridge: MIT Press, 1987), esp. chap. 8. For Bernstein's argument, see *The New Constellation*, chap. 7.

4. We have our own reservations about the extreme to which postmodernist claims are sometimes taken: for instance, when claims are made that there is no objective reality knowable under any circumstances. We disagree with such arguments, yet we would argue that in the case of contested political issues the postmodernist insight that there cannot be any single perspective that is objectively correct is an important one, especially when it is used to point to the political power inherent in the assertion that there is such a perspective.

5. Richard J. Bernstein, *Beyond Objectivism and Relativism* (Philadelphia: University of Pennsylvania Press, 1985), 8.

6. David Collingridge and Colin Reeve, *Science Speaks to Power* (New York: St. Martin's Press, 1986).

7. Thomas S. Kuhn, *The Structure of Scientific Revolutions* (Chicago: University of Chicago Press, 1970).

8. This view is closely associated with the work of Robert Merton on the sociology of science. For a recent application of this work to the social sciences, see Paul Diesing, *How Does Social Science Work?* (Pittsburgh: Pittsburgh University Press, 1991).

9. Helen Longino, *Science as Social Knowledge* (Princeton: Princeton University Press, 1990).

10. Collingridge and Reeve, *Science Speaks to Power*.

11. An interesting example of the tension between the needs of policymakers and the procedures of scientists is provided by Stephen J. Gould's analysis of congressional hearings on

fraud by researchers receiving federal grants. Gould argues that Congress expects certainty and correct, usable knowledge to result from the scientific research paid for by the federal government. When researchers fall short of this goal, they are often accused of incompetence or, worse still, fraud. Gould agrees that it is important to weed out fraud and incompetence; however, he believes that they are relatively small problems. The real problem is that federal policymakers often confuse incompetence or fraud with error. It is inevitable that most research will, in the long or short run, prove wrong. Indeed, scientific "progress" depends on error. Failing to understand this, congressmen accuse erroneous research of being the result of either incompetence or fraud. To the extent that it raises the costs of error, the ultimate result of this attempt to closely police research for fraud may be poorer science. Stephen Jay Gould, "Judging the Perils of Official Hostility to Scientific Error," *New York Times,* July 30, 1989, sec. 4, p. 6.

12. Collingridge and Reeve, *Science Speaks to Power.*

13. On this point, see especially Christopher Lasch, *The True and Only Heaven: Progress and Its Critics* (New York: W. W. Norton, 1990).

14. This view of public education goes back to the origins of the managerial discourse. See, for example, Woodrow Wilson's discussion in "The Study of Administration," in *Classics of Public Administration,* ed. Jay M. Shafritz and Albert C. Hyde (Chicago: Dorsey Press, 1987).

15. Collingridge and Reeve, *Science Speaks to Power.*

16. As Bernstein puts it, "The relativist not only denies the positive claims of the objectivist but goes further. In its strongest form, relativism is the basic conviction that when we turn to the examination of those concepts that philosophers have taken to be the most fundamental— whether it is the concept of rationality, truth, reality, right, the good, or norms—we are forced to recognize that in the final analysis all such concepts must be understood as relative to a specific conceptual scheme, theoretical framework, paradigm, form of life, society, or culture" (*Beyond Objectivism and Relativism,* 8).

17. In our analysis of hazardous waste regulation in New Jersey, for example, we found that a key problem was the pluralist assumption

by policymakers that guaranteeing membership for *organized* interests on decision-making bodies would result in adequate representation for all affected interests. However, structural barriers to organization meant that many affected interests were unorganized and, thus, unrepresented. When these interests were later mobilized, they challenged the fairness of the entire decision-making process (see chapter 6).

18. As Alford and Friedland put it, in pluralism "the state remains implicit." *Powers of Theory,* xx.

19. As we noted in chapter 2, it is the ultimate belief that self-interest, correctly harnessed, provides the best engine for achieving a social good that makes pluralism the political analogue to a faith in the efficacy of private markets.

20. John Dewey, *The Public and Its Problems* (Chicago: Swallow Press, 1927), 131. Dewey's argument is rooted in his pragmatic philosophy, which, we argue below, uses the political vocabulary of the communitarian discourse.

21. Thomas McCollough, *Moral Imagination and Public Life* (Chatham, N.J.: Chatham House, 1991), 72.

22. See especially Dahl's *Democracy and Its Critics* (New Haven: Yale University Press, 1989).

23. Benjamin Ginsberg and Martin Schefter provide an example of the significance of changes in the groups that citizens choose to identify with in their analysis of Republican electoral success in the 1980s. They argue that "Republicans have worked to reshape the political attachments of business executives [unifying big and small businesspeople], middle class suburbanites [switching their self-identification from beneficiaries to taxpayers], blue collar ethnics [transforming them from workers to patriots], and white southerners [shifting their self-perception from southerners to evangelicals]." *Politics by Other Means* (New York: Basic Books, 1990), 108, and chap. 4 more generally.

24. Ira Katznelson provides the best discussion of the separation of workplace and residence and its implications for political activity: *City Trenches* (Chicago: University of Chicago Press, 1981). We explore the differences for social regulation when the focus of community action is organized around residence as opposed

to workplace (see chapter 4 and the case studies in chapter 6).

25. See especially Dewey's *The Public and Its Problems* and *Reconstruction in Philosophy* (New York: Mentor Books, 1920). We also rely on more recent analysts of Dewey's work: James Carey, Robert Westbrook, and Richard Rorty. Although they tend to agree about the implications of Dewey's pragmatism for a theory of knowledge that avoids the pitfalls of objectivism and relativism, they vehemently disagree over Dewey's conclusions about the implications of this perspective for democratic politics. Our own perspective, outlined in our dialogic model, is much closer to Carey's and Westbrook's than to Rorty's views on Dewey's political theory. See James Carey, *Communications as Culture* (Boston: Unwin Hyman, 1989); Robert B. Westbrook, *John Dewey and American Democracy* (Ithaca: Cornell University Press, 1991), and Rorty, *Objectivity, Relativism, and Truth*.

26. *The Public and Its Problems*, esp. chaps. 5, 6, and also Westbrook, *John Dewey*.

27. Bernstein, *Beyond Objectivism and Relativism*, 172.

28. This view is consistent with Dewy's interpretation of the development of philosophy, expressed in his *Reconstruction in Philosophy*.

29. For an excellent example of the role that dialogue plays in achieving consensus, see the description of the face-to-face debate at professional meetings between advocates of different models of subatomic physics in the early twentieth century in Richard Rhodes, *The Making of the Atomic Bomb* (New York: Simon and Schuster, 1986).

30. Kuhn argues that the presence of subjective values does not necessarily make the outcome of discourse an arbitrary or relativistic process. His argument is summarized by Bernstein, *Beyond Objectivism and Relativism*, 56: "If I report that I like something (Kuhn's example is a film), short of claiming that I am lying or somehow deceiving myself no one can disagree with my subjective report. But if I claim that the film was a 'terrible potboiler,' then I am making a *judgement* which is discussable. If I am challenged in making this judgement, then I am expected to give *reasons* that support my judgement. According to Kuhn, the deliberations concerning the choices of [paradigms] have this

judgmental character. They are 'eminently discussable, and the man who refuses to discuss his own cannot expect to be taken seriously.' . . . The reasons do not 'prove' the judgement, they support it. In a concrete situation there can be better and worse reasons (even though there are no clear and precise rules for sorting out what is better and worse)."

31. Cited in Rhodes, *The Making of the Atomic Bomb*, 77 (emphasis in original).

32. Joseph Rouse, *Knowledge and Power* (Ithaca: Cornell University Press, 1987), 7.

33. This critique of liberal and conservative views of community draws heavily from Benjamin R. Barber, *Strong Democracy* (Berkeley: University of California Press, 1984), chap. 9. For a similar positive view of communities, see also Lasch, *The True and Only Heaven*.

34. In *The Moral Imagination and Public Life*, McCollough points out that liberal social theory is flawed because it sees nothing standing between the individual and the society as a whole, ignoring the role of older mediating forms of social organization (e.g., community, family, religion, etc.), which have been threatened by modernization but not destroyed.

35. For an argument that links the increasing number of zero-sum decisions to the breakdown of consensus, increasing participation and the "overload of democracy," see Douglas Yates, *The Ungovernable City* (Cambridge: MIT Press, 1977). For a more general argument about the inability of pluralist politics to deal with zero-sum decisions, see Lester Thurow, *The Zero-Sum Society* (New York: Basic Books, 1980).

36. Lasch, *The True and Only Heaven*, chap. 5.

37. James Morone, *The Democratic Wish* (Basic Books, 1990), esp. Conclusion.

38. Rhodes, *The Making of the Atomic Bomb*, cites many vivid examples of the importance for the development of nuclear physics of replicable anomalous experimental findings. In the 1930s, for example, accepted theories of subatomic structure held that fission (the splitting of an atom's nucleus) was impossible. Experiments involving the neutron bombardment of uranium by Irène and Frédéric Joliot-Curie seemed to indicate that such fission was occurring. These findings were, at first, discarded by the physics community as the result of imper-

fect technique. However, when these results were replicated by several other laboratories, they came to be accepted, and theories of subatomic structure were modified accordingly. Central to the modification of such theories was the ability of scientists throughout the community to verify the seemingly improbable findings of the French researchers.

39. Indeed, a common complaint in the social sciences is that consensus is rarely achieved because research findings are seldom reproduced by different scholars.

40. In *The Mismeasure of Man* (New York: W. W. Norton, 1981), Stephen J. Gould illustrates this sort of pull when he describes how, in his attempt to refute empirically the dubious claims of nineteenth-century "craniometry" about racial differences in intellectual abilities based upon brain size, he consistently mismeasured the cranial capacity of the skulls he examined. These mismeasurements always were in the direction of supporting Gould's arguments. The same sort of systematic pressures may have been operating on researchers at the University of Utah in their rapidly discredited attempts to demonstrate cold fusion.

41. Indeed, some critics of the current organization of the scientific community argue that because research is concentrated in the hands of scientists at major universities necessarily supportive of the status quo, it does deviate significantly from the standards of undistorted communication and scientific rationality. See David Dickson, *The New Politics of Science* (New York: Pantheon, 1984). David Noble makes a similar case, arguing that the development of the engineering profession in the United States was seriously distorted (especially with respect to the absence of consideration by engineers of the public good, as opposed to private profit) by its dependence upon corporate capitalism. *America By Design* (New York: Oxford University Press, 1977). Longino also supports this position, arguing that a scientific community is protected from biases only to the degree that it is open to those who can challenge the claims and underlying assumptions of others in the community. *Science as Social Knowledge.*

42. Cited in Neil Postman, *Amusing Ourselves to Death* (New York: Penguin Books, 1985), 108.

43. Establishing the legitimacy of the decision-making process is a crucial task in democratic political communities. One key to legitimacy may be coming to view any decision about the substantive goals of the community as tentative and evolving, rather than definitive. This allows, through ongoing and repeated interaction, the buildup of trust and a reduction in the stakes of any single episode of decision-making (see below).

44. John Forester, "The Policy Analysis–Critical Theory Affair: Wildavsky and Habermas as Bedfellows?" in John Forester, ed., *Critical Theory and Public Life* (Cambridge: MIT Press, 1985).

45. Ibid., 274.

46. We address the impact of NIOSH's findings in chapter 7. For a critique of the ways in which the OSHA legislation was implemented, see Charles Noble, *Liberalism at Work* (Philadelphia: Temple University Press, 1986).

47. For Dewey's approach to the relation between science and democracy, especially his identification of access to knowledge as a prime political resource, see *The Public and Its Problems*, 174–81. For interpretations of the connection between democracy and science in Dewey's thought more generally, see Westbrook, *John Dewey*; Lasch, *The True and Only Heaven*; and Carey, *Communication and Culture.*

48. On these parallels, see Bernstein, *Beyond Objectivism and Relativism.*

49. We do not claim to be experts in the complex, subtle, and sometimes confusing (at least to us) arguments of Habermas. We base our arguments on our reading of selected works by Habermas that deal with the specific issues of concern to us. For an overview of Habermas's entire project, we rely on the more comprehensive interpretations of Steven White, *Jürgen Habermas* (New York: Cambridge University Press, 1988); David Ingram, *Habermas and the Dialectic of Reason* (New Haven: Yale University Press, 1987); Thomas McCarthy, *The Critical Theory of Jürgen Habermas* (Cambridge: MIT Press, 1978).

50. Habermas himself recognizes the strong connections between his work and the work of American pragmatists, especially Dewey. See Bernstein, *The New Constellation*, chap. 7, esp. 202–08.

51. Although Habermas does not phrase his

argument specifically in terms of policy analysis, his critique of technical rationality and its tendency to dominate other forms of rationality (its "colonization of the life-world," in his language) anticipates such an argument. Jürgen Habermas, *Toward a Rational Society* (Boston: Beacon Press, 1970). For an excellent discussion of the differences between mainstream and critical policy analysis, see Forester, "Policy Analysis-Critical Theory."

52. Habermas, *Rational Society*. See also Ingram, *Dialectic of Reason*, chap. 2.

53. Habermas has devoted a great deal of effort to defining exactly what is entailed in undistorted communication. See especially his *Knowledge and Human Interest* (Boston: Beacon Press, 1971); *Theory and Practice* (Boston: Beacon Press, 1973). Our simplification of these efforts to four claims relies on Brian Fay, *Critical Social Science* (Ithaca: Cornell University Press, 1987), and Forester, "Policy Analysis-Critical Theory."

54. Jane J. Mansbridge, *Beyond Adversarial Democracy* (Chicago: University of Chicago Press, 1980).

55. Unitary democracy shares some assumptions with the managerial discourse. It is similar because both unitary and managerial perspectives assume the existence of a discoverable public interest and that impediments to its discovery involve inadequate information. They differ significantly, however, because unitary models assume that democratic participation is possible and necessary to legitimate the outcomes of decisions. Whether an issue is technical or political is not an objective characteristic of the issue itself and cannot be decided outside the operation of democratic politics. Thus, the very decision to treat an issue as unitary is a political one that must be made in a democratic fashion requiring the active participation of citizens. In contrast, managerial approaches usually ignore this stage in the decision-making process or assume that it too is the province of experts.

56. Richard Bernstein makes a similar point: "What we desperately need today is to learn to think and act more like the fox than the hedgehog—to seize upon those experiences and struggles in which there are still the glimmerings of solidarity and the promise of dialogical communities in which there can be genuine

mutual participation and where reciprocal wooing and persuasion can prevail." *Beyond Objectivism and Relativism*, 228.

57. Robert Axelrod, *The Evolution of Cooperation* (New York: Basic Books, 1984). For interesting applications of Axelrod's ideas to environmental policy-making, see Daniel Mazmanian and David Morell, "The Elusive Pursuit of Toxics Management," *The Public Interest* 90 (Winter 1988), and Barry G. Rabe, "Hazardous Waste Facility Siting: Subnational Policy in Canada and the United States," paper presented at the American Political Science Association Meetings, 1989.

58. Barber, *Strong Democracy*.

59. Ibid., 132 (emphasis deleted).

60. Barber would raise these sorts of objections about Florida's Amnesty Days program and others like it, which attempt to limit participation to divisive issues that elites have failed to solve (see chapter 6).

61. Ibid., 234.

62. Ibid., 191.

63. This process is especially well illustrated by hazardous waste regulation in New Jersey, where media coverage of dramatic hazardous waste disasters pushed the issue to the top of the list of citizens' concerns as revealed in public opinion polls. These polls, in turn, became the subject of news stories that reinforced public concern and focused the attention of politicians on the issue (see chaps. 1, 5).

64. For a fuller discussion of these issues, see Michael X. Delli Carpini and Bruce A. Williams, "'Fictional' and 'Non-fictional' Television Celebrate Earth Day, or Politics is Comedy Plus Pretense," *Cultural Studies* 8 (January 1994), 74–98; and id., "Methods, Metaphors, and Media Messages: The Uses of Television in Political Conversations," *Communication Research* 21 (December 1994), 782–812.

65. Barber presents a detailed plan for creating strong democracy that we return to in chapter 8.

4 / Democracy, Redistribution, and Social Regulation in a Federal System

1. Daniel Press, *Democratic Dilemmas in the Age of Ecology* (Durham: Duke University Press, 1994).

2. One of the best examples of the inability of local communities to address such issues on their own is the persistent siting of dangerous industrial and storage facilities in or near poor black neighborhoods, exposing the residents to much higher levels of toxic pollution than is the case in more affluent white neighborhoods. For a general discussion of what is termed environmental racism, see Robert Bullard, *Dumping in Dixie: Race, Class and Environmental Quality* (Boulder: Westview Press 1990). For a heartbreaking description of toxic pollution in the poor black city of East St. Louis and its effects on children, see Jonathan Kozol, *Savage Inequalities* (New York: Crown Publishers, 1991), chap. 1. See also our discussion of this issue in chapter 7.

3. See our discussion in chapter 2. On the redistributive effects of all forms of regulation, see also Lester Thurow, *The Zero-Sum Society* (New York: Basic Books, 1980).

4. See Eugene Bardach and Robert A. Kagan, *Going by the Book* (Philadelphia: Temple University Press, 1982).

5. See Robert L. Bish, *The Political Economy of Metropolitan Areas* (Chicago: Markham Books, 1971); Charles Tiebout, "A Pure Theory of Local Expenditures," *Journal of Political Economy* 4 (October 1956); Vincent Ostrom, *The Intellectual Crisis in Public Administration* (University: University of Alabama Press, 1965).

6. For example, they have opposed reforms aimed at combining localities into larger metropolitan governments. See Ostrum, *Intellectual Crisis*, and *Understanding Local Government*.

7. David Lowery and William E. Lyons, "Citizen Responses to Dissatisfaction in Urban Communities: A Partial Test of a General Model," *Journal of Politics* 51 (November 1989), 841–68.

8. Paul E. Peterson, *City Limits* (Chicago: University of Chicago Press, 1981).

9. Although Peterson focuses primarily upon city politics, he notes that his arguments also apply in large measure to the dynamics of state policy-making. Indeed, he tests his theory on state expenditure data.

10. For a discussion of how this dynamic works to the advantage of automobile corporations by fostering an increasingly lucrative government benefit package as states compete

to attract new factories, see Bruce A. Williams, "Regulation and Economic Development," in *Politics in the American States*, ed. Virginia Gray, Herbert Jacob, and Robert Albritton (Harper/Collins, 1990).

11. Peterson, *City Limits*, 41.

12. Ibid., 43.

13. Ibid., 44.

14. It should be noted that Peterson's definition of redistribution is more restrictive than the one we have been employing up until now. He defines redistribution as only those policies that shift income from high-income to low-income residents, while we have defined redistribution as any policy that shifts income from one group to another. In our empirical analysis in chapter 5 we adopt the more restrictive definition by examining the impact on regulatory policy of the economic significance (and thus contribution to a state's tax base) of those industries adversely affected by regulation.

15. Charles Lindblom, "The Market as Prison," *Journal of Politics* 4 (May 1982).

16. This failure may be primarily the result of his main focus on the dynamics of local politics. He does not examine the ways in which national policy-making is used to pursue the redistributive policies slighted at the state and local levels. However, it also results from his argument that in the end all citizens, regardless of class background, share a common interest in the expansion of the local economy and hence developmental policies. Thus, he concludes that even if their specific demands for redistribution are ignored, the interests of the poor are actually served by this, because they too benefit from the success of localities in the competition for economic development that would be thwarted by responding to demands for redistribution.

17. Ira Katznelson, *City Trenches* (Chicago: University of Chicago Press, 1981).

18. Peterson, *City Limits*, 89.

19. Katznelson, *City Trenches*, 37.

20. On this subject, see also David Montgomery, *The Fall of the House of Labor: The Workplace, the State, and American Labor Activism, 1865–1925* (New York: Cambridge University Press, 1981).

21. Our discussion of the impact of the Progressive movement in chapter 2 focused mainly on its effect upon the ideology of federal regu-

lation. Of course the Progressives also sought reform in urban government, and the impact of that reform was to limit even more the range of pluralist politics at the local level. But that movement's reliance on the managerial discourse had consequences for labor-management relations as well. In the workplace, "scientific management" forced organized labor into even more narrow pursuits, by tying wages to efficiency. Control of the workplace became purely the province of management, and management the province of experts, as detailed by Montgomery, *Fall of the House of Labor:* "The essence of scientific management was a systematic separation of the mental component of commodity production from the manual. The functions of thinking and deciding were what management sought to wrest from the worker, so that the manual efforts of wage earners might be directed in detail by a 'superior intelligence.' . . . Workers' happiness would come through an abundance of material goods, and that abundance was to be created through ever increasing productivity" (127).

22. Ibid.

23. Peterson, *City Limits,* 108.

24. We explore these two types of distortion more completely in chapter 7.

25. See note 1.

26. Richard A. Harris and Sidney M. Milkis, *The Politics of Regulatory Change: A Tale of Two Agencies* (New York: Oxford University Press, 1989), 274.

27. See our discussion, chapter 3, and also John Forester, "The Policy Analysis-Critical Theory Affair: Wildavsky and Habermas as Bedfellows?" in John Forester, ed. *Critical Theory and Public Life* (Cambridge: MIT Press, 1985).

28. Ibid., 274.

29. We do not treat comprehensively the struggle over voting rights, CAPs, or LSC. Rather, these examples are suggestive, and we use them to demonstrate the degree to which precedents exist for the sort of federal efforts we outline in this chapter. We also select these examples to reinforce our more general argument that the problems of social regulation at the state and local levels raise profound questions of citizenship that can be answered only if we move beyond the boundaries implicitly adopted by most studies of the new social regulation.

30. Abigail Thernstrom, *Whose Votes Count?* (Cambridge: Harvard University Press, 1987).

31. For a representative negative treatment, see Charles Murray, *Losing Ground* (New York: Basic Books, 1984). For a generally positive assessment of the War on Poverty, see John E. Shwarz, *America's Hidden Success* (New York: W. W. Norton, 1983).

32. For treatments of the CAP legacy, see Paul Peterson, Barry Rabe, and Kenneth Wong, *When Federalism Works* (Washington: Brookings Institution, 1986); Douglas Yates, *The Ungovernable City* (Cambridge: MIT Press, 1975).

33. Experimental research, which we discuss more fully in chapter 6, suggests that citizens are reassured about hazardous waste facilities when they are guaranteed an ongoing role in their management. Michael L. Poirier Elliott, "Improving Community Acceptance of Hazardous Waste Facilities through Alternative Systems for Mitigating and Managing Risk," *Hazardous Waste* 1 (1984), 397–410.

34. Mark Kessler, *Legal Services for the Poor* (New York: Greenwood Press, 1987), 6–7.

35. Figures from ibid., 8.

36. Indeed, this is one of the goals of several grassroots groups we discuss in chapter 7.

37. Again, in chapter 7 we discuss the important but inevitably limited attempts of grassroots environmental groups to build these linkages.

5 / The Politics of Hazardous Waste Regulation

1. For a description of events at Love Canal, see Michael H. Brown, *Laying Waste: The Poisoning of America by Toxic Chemicals* (New York: Pantheon, 1980).

2. One metric ton equals 2,205 pounds.

3. Mary Worobec, "An Analysis of the Resource Conservation and Recovery Act," *Environment Reporter* (August 22, 1980), 633–46.

4. The earlier estimate by the EPA placed hazardous waste generation at about 350 pounds per U.S. inhabitant, while the later estimate reached nearly 2,500 pounds per U.S. inhabitant.

5. GAO, *Superfund: Extent of Nation's Potential Hazardous Waste Problem Still Unknown,*

RCED-88–44 (Gaithersburg, Md.: General Accounting Office, 1988).

6. On this point, see generally Samuel S. Epstein et al., *Hazardous Waste in America* (San Francisco: Sierra Club, 1983).

7. Congressional Reference Service, *Resource Issues from Surface, Groundwater and Atmospheric Contamination* (Washington, D.C.: Government Printing Office, 1980), 75–76.

8. Mining, petroleum, and metallurgical operations have always produced a substantial amount of hazardous waste and have received specialized attention through the years as hazardous industries. Unlike those of the petrochemical industry, the hazards of mining, refining, and smelting ore are comparatively observable. RCRA defines hazardous waste generally as a solid waste posing a substantial hazard to health or the environment when improperly managed. The EPA promulgated regulations in 1980 in response to that vague definition in ways not intuitively obvious:

1. ignitability—posing a fire hazard during routine management;

2. corrosivity—ability to corrode standard containers or to dissolve toxic components of other wastes;

3. reactivity—tendency to explode under normal management conditions, to react violently when mixed with water, or to generate toxic gases; and

4. "extraction procedure" toxicity—the presence of certain toxic materials at levels greater than those specified in the regulation, with reference to a list of designated hazardous waste meeting the "general toxicity" criteria below:

a. the degree of toxicity of the toxic constituents in the waste;

b. the concentration of these constituents in the waste;

c. the potential for these constituents or their by-products to migrate from the waste into the environment;

d. the persistence and degradation potential of the constituents or their toxic by-products in the environment;

e. the potential for the constituents or their toxic by-products to bioaccumulate in ecosystems;

f. the plausible and possible types of im-

proper management to which the waste may be subjected;

g. the quantities of the waste generated; and

h. the record of human health and environmental damage that has resulted from past improper management of wastes containing the same toxic constituents.

Environment Reporter (August 22, 1980), 639–40.

9. According to the EPA, the chemical industry is responsible for an estimated 79.3 percent of the hazardous wastes generated in America. Although chemical manufacturers, in most cases, now store, recycle, treat, and dispose of their hazardous wastes on-site, at the point of generation, using sophisticated technologies to minimize health risks, the products manufactured often become hazardous waste after being sold and used by other businesses and consumers.

10. Barry Commoner, *The Closing Circle* (New York: Alfred A. Knopf, 1971).

11. Market incentives, without government supervision, made land disposal by far the most feasible economically. And there was no reason in the short run to pursue other alternatives, for example, source reduction, recycling, or incineration. In 1980, land disposal was cheaper than any other method by a factor of between 2.5 and 1,000. EPA, *Hazardous Waste Generation and Commercial Hazardous Waste Management Capacity: An Assessment* (Washington, D.C.: Government Printing Office, 1980), 15.

12. Worobec, "Analysis," 634, quoting Thomas F. Williams, then-deputy director of the EPA Office of Public Awareness, in a 1980 speech recalling the early days in the EPA.

13. See note 3, above.

14. On this point, see Epstein et al. *Hazardous Waste,* 189–97.

15. Ibid., 192.

16. Ibid., 192–93. Forewarned of the prospect of tougher regulation without any immediate effect on existing disposal practices, hazardous waste generators had a great incentive to dispose of their hazardous waste stores as quickly as possible without fear of penalty. As it turned out, they had plenty of time!

17. Ibid., 189–91.

18. Douglas M. Costle, then-administrator

of the EPA, stated in October 1979 that the agency had received twelve hundred sets of comments on its proposed program, forming a stack seven feet high. Most of the comments were industry attempts to lay groundwork for subsequent suits under administrative law, once the regulations went into effect. Even the Carter administration's Regulatory Analysis Review Group attacked the proposed program: see Worobec, "Analysis," 636–38.

19. U.S. House of Representatives. *Hazardous Waste Disposal, Report from House Subcommittee on Oversight and Investigations,* print 96-IFC 31 (Washington, D.C.: Government Printing Office, 1979).

20. For example, the EPA was forced to abandon its originally proposed one-hundred-kilogram (kg) exemption for small generators for a one-thousand-kg exemption. Information developed later showed that most of the hazardous waste posing problems came from the smaller generators. Lack of information on hazardous waste disposal technology similarly led both Congress and the EPA early on to encourage land-filling rather than other methods of hazardous waste disposal. Not until November 1984, with the passage of the Hazardous and Solid Waste Amendments to RCRA, were these misperceptions officially recognized and corrected.

21. Eckardt C. Beck, quoted in Steven Cohen and Marc Tipermas, "Superfund: Pre-implementation Planning and Bureaucratic Politics," in *The Politics of Hazardous Waste Management,* ed. James P. Lester and Ann O'M. Bowman (Durham: Duke University Press, 1983), 44.

22. Robert A. Roland, president of the Chemical Manufacturing Association and leading industry lobbyist during the debate over CERCLA, stated, "The solid waste disposal problem, including toxic or hazardous wastes, is not just the problem of the chemical industry. It is a result of society's advanced technology and pursuit of an increasingly complex lifestyle. . . . Everyone should realize that the blame does not belong to a single company, or a single industry, but to all of us as individuals and as an advanced society. Rather than looking for scapegoats, we should recognize the dilemma and consider new ways to encourage the disclosure

of dump site information and ways to limit the crushing liabilities that could result." Quoted in Epstein et al., *Hazardous Waste,* 206.

23. For a detailed description of congressional and interest-group machinations over CERCLA, see ibid., 197–222.

24. In a dramatic lesson about the dangers of entrepreneurship, Culver was defeated in his 1980 reelection bid by John Grassley. According to Elizabeth Drew, in retribution for Culver's sponsorship of the redistributively financed Superfund, Political Action Committees representing the chemical and petroleum industries made substantial and decisive campaign contributions to Grassley: see her *Politics and Money* (New York: Macmillan, 1983).

25. The joint and several liability and strict liability features of CERCLA were struck from the bill before passage in the Senate to avoid a filibuster by Sen. Jesse Helms (R-NC). The size of the Superfund, which had vacillated during congressional deliberations between $1 and $4 billion, was finally set at $1.6 billion, with $1.38 billion to come from a tax on chemical feedstocks and crude oil and $220 million from a congressional appropriation. Oil spills were deleted from coverage in the final legislation, as were provisions for victim compensation. Interestingly, the first and most influential judicial gloss on the liability provisions of CERCLA essentially gave a "joint and several liability" reading to the law, thus providing the EPA with a key enforcement tool that has come to dwarf the other features of CERCLA in terms of its potential redistributive impact on industry. See *U.S. v. Chem-Dyne Corp.,* 572 F. Supp. 802 (1983).

26. Cohen and Tipermas, "Superfund," 56.

27. Two insiders in the EPA's preimplementation strategy reported that the Office of Hazardous Emergency Response, the "Superfund Office," had already established a softball team and printed T-shirts with a Superfund logo before Carter ordered the implementation of CERCLA. Ibid., 56.

28. Richard Riley, "Toxic Substances, Hazardous Wastes, and Public Policy: Problems in Implementation," *The Politics of Hazardous Waste Management,* ed. Lester and Bowman (Durham: Duke University Press, 1983), 24–42.

29. Epstein et al., *Hazardous Waste*, 213–14.

30. Cohen and Tipermas, "Superfund," 57.

31. Robert C. Mitchell, "Public Opinion and Environmental Politics in the 1970s and 1980s," in *Environmental Policy in the 1980s: Reagan's New Agenda*, ed. N. J. Vig and M. E. Kraft (Washington, D.C.: Congressional Quarterly Press, 1984), 51–74.

32. Gorsuch was a conservative lawyer and former state legislator from Colorado who, just months before her appointment, had sought to minimize her state's hazardous waste program (see, *Inside EPA*, March 20, 1981, SV 2 #12, p. 8); but she was a close ally of Reagan's new secretary of the interior, James Watt. Watt represented the conservative agenda of the Heritage Foundation, a Washington-based research group founded (and funded) in part by the Colorado brewery magnate Joseph Coors. As a primary architect of the so-called sagebrush rebellion, Watt called for an end to EPA involvement in land management decisions in the mountain and western states. His influence in the Reagan administration's early approach to environmental regulation overwhelmed more moderate voices during the transition period. See Michael J. Kraft, "A New Environmental Policy Agenda: The 1980 Presidential Campaign and Its Aftermath," in *Environmental Policy in the 1980s*, ed. Vig and Kraft, 38–40.

33. On October 23, 1981, when the EPA issued its Interim National Priority List of Superfund sites, officials commented that Superfund money would be needed to clean up only 170 of the expected 400 sites on the final NPL. The cost of each cleanup was estimated to be only $4 million per site, well within the $1.6 billion originally allocated under CERCLA: *Environment Reporter* (October 30, 1981), 807.

34. Richard N. L. Andrews, "Deregulation: The Failure at EPA," in *Environmental Policy in the 1980s*, ed. Vig and Kraft, 169.

35. *Environment Reporter*, (October 30, 1981), 811.

36. Walter A. Rosenbaum, "The Politics of Public Participation in Hazardous Waste Management," in *The Politics of Hazardous Waste Management*, ed. Lester and Bowman, 187–89.

37. See Steven Cohen, "Defusing the Toxic Time Bomb: Federal Hazardous Waste Programs," in *Environmental Policy in the 1980s*, ed. Vig and Kraft, 273–91, for a succinct and accurate description of the events surrounding both RCRA and CERCLA implementation during the late Carter and early Reagan administrations. He refers to the original "shovels first, and lawyers later" intent of CERCLA as being transformed into a "lunch now, lawyers maybe, but shovels never" attitude!

38. Statement by Rita Lavelle, quoted in ibid., 285.

39. She had since married Robert Burford and changed her name to Anne M. Burford.

40. Cohen, "Defusing the Toxic Time Bomb," 285.

41. J. Wessinger, quoted in *Environment Reporter* (March 9, 1984).

42. Florio, quoted in *Environment Reporter* (September 28, 1984), 838. The National Wildlife Federation made a similar plea to Ruckelshaus in a letter dated August 31, 1984: "There are suspicions in certain quarters that there may be a conscious strategy on the part of the Administration to stall any NPL update as long as possible (perhaps until after the November election) in order to minimize any statutory impact on the superfund amendments now pending in Senate Committees. . . . [Ruckelshaus was urged] to resist any effort to politicize the NPL process." Quoted in *Environment Reporter* (September 7, 1984), 722.

43. The legislation phased out land-filling as a discredited and unsafe technology for hazardous waste disposal, imposed RCRA regulations on small-quantity (one hundred kg) hazardous waste generators, provided for underground storage tank regulation, authorized appointment of an ombudsman within EPA to foster public input and set performance standards and multiple deadlines aimed at reducing EPA discretion to avoid the intent of the amendments. *Environment Reporter* (October 26, 1984), 1136–42.

44. *Environment Reporter* (May 23, 1986), 85.

45. Office of Technology Assessment, *Superfund Strategy* (Washington, D.C.: Government Printing Office, 1985). These criticisms surfaced in the states, too, as we found in our case study of Florida, where serious questions were raised about the accuracy of NPL rankings of sites in the state (see note 81, below).

46. *Environment Reporter* (January 22, 1988), 2043.

47. EPA Journal, 1987, 16.

48. Office of Technology Assessment, *Super-fund*.

49. Winston J. Porter, "Editorial," *Washington Post* (July 2, 1988), 17A.

50. Adam Clymer, "Polls Contrast U.S.'s and Public's Views," *New York Times* (May 22, 1989), 11A.

51. This point is tellingly made in Marc K. Landy, Marc J. Roberts, and Stephen R. Thomas, *The Environmental Protection Agency: Asking the Wrong Questions* (New York: Oxford University Press, 1990).

52. For example, Murray Weidenbaum, *The Costs of Government Regulation of Business* (Washington, D.C.: Government Printing Office, 1978); Murray Weidenbaum and Robert DeFina, *The Cost of Federal Regulation of Economic Activity*, Reprint no. 88 (Washington, D.C.: American Enterprise Institute, 1978). The conservative criticism of the new social regulation can be contrasted with the more widely held view that many types of older, economic regulation were less efficient than no regulation at all. The discussion of deregulation is covered extensively in chapter 2.

53. See our discussion of this issue in chapter 2.

54. Numbers 11,821 and 12,044, respectively.

55. A good example of this is contained in a memorandum of May 1978 from a regional EPA administrator rebuking the EPA's Hazardous Waste Division for attempting to invoke RCRA's "imminent hazard" provision: "As we are all aware, hazardous waste management facilities are inherently hazardous. Determination of imminent hazard is, in part, a legal matter, and must in my view involve a risk of significant magnitude to warrant federal intrusion into an area that has historically been handled by the state and local sector." Cited in Epstein et al., *Hazardous Waste*, 228.

56. Richard A. Harris and Sidney M. Milkis, *The Politics of Regulatory Change: A Tale of Two Agencies* (New York: Oxford University Press, 1989), 275.

57. Cohen, "Toxic Time Bomb."

58. Such conscious manipulation of the languages of regulatory legitimacy would, however, be consistent with the cynical orchestration of both econometric models and the rhetoric of supply-side economics revealed in William Greider, "The Education of David Stockman," *The Atlantic* (December 1981).

59. For a full treatment of these issues, see Bruce A. Williams and Albert R. Matheny, "Testing Theories of Social Regulation: Hazardous Waste Regulation in the American States," *Journal of Politics* 46 (1984), 428–58. To summarize, our investigation of specific state hazardous waste programs indicates that attempting to isolate hazardous waste regulation expenditures in 1980 would have yielded very unstable figures, not truly representative of a state's actual commitment to hazardous waste regulation. In many cases in 1980, state expenditures on hazardous waste had not been differentiated from general environmental expenditures, and, in any case, the impact of a state's effort at hazardous waste regulation would be directly affected by other, more general efforts at land and water quality monitoring. One statistical implication of using this measure should be noted. It is likely that not all states allocate the same proportion of their total resources in this area (i.e., land and water quality control) to efforts that are related to hazardous waste. Thus, the percentage of such funds that reflect actual effort in the area of hazardous waste regulation may differ from state to state. We assume, however, that while this percentage may vary, it varies in a random, unbiased way from state to state. The result of such measurement error will be an attenuation of the strength of relations produced by our regression equations. See Jim C. Nunnally, *Psychometric Theory* (New York: McGraw-Hill, 1967), chap. 6. Given the data available for 1980 in this area, we kept this possibility in mind as we interpreted our data analysis.

60. C. K. Rowland and Richard Feiock, "Environmental Regulation and Economic Development: The Movement of Chemical Production among the States," *Western Political Quarterly* 43 (September 1990), 561–76. This study, supportive of our analysis and assumptions, was conducted after state hazardous waste expenditure data became available later in the decade.

61. We began the fieldwork for our case studies in 1983, when hazardous waste regulation had already become a prominent feature of all three states' policy agendas. We spent several weeks in each state visiting communities threat-

ened by leaking waste sites, touring various hazardous waste facilities, examining media accounts of the issue's development, and reading available public documents. Our primary task, however, was identifying and interviewing the key players in this policy area: bureaucrats, legislators and their staffs, representatives of industry trade associations, leaders of state and local environmental and citizen groups, and journalists who covered environmental issues. Interviews ranged from one to three hours and involved both open-ended and closed-ended questions. Following these initial interviews and site visits, we maintained contact with our informants over the next eight years through written correspondence and telephone calls as well as occasional visits. Getting to know the public and private officials and citizens involved in this area of social regulation allowed us to see, over an extended period of time, how these individuals articulated their understanding of the daunting problems posed by hazardous waste regulation.

62. It has been estimated that the state produces 8 percent of all the hazardous wastes generated in the nation. David Morell and Christopher Magorian, *Siting Hazardous Waste Facilities* (New York: Ballinger, 1982), 10.

63. Ibid.

64. Compare these figures with the sluggish performance of Ohio's GSP over the same time period: 1980–84: 6.8 percent; 1984–85: 5.7 percent; 1985–86: 5.0 percent. Florida's performance was roughly comparable with New Jersey's during this period. Unemployment in Ohio and Florida increased between 1980 and 1984 from 8.4 percent to 9.4 percent and from 5.9 percent to 6.3 percent, respectively. In 1987, Ohio's unemployment rate was 7.0 percent and Florida's was 5.3 percent. Ohio actually slipped in its wealth ranking from twenty-fourth to twenty-fifth in 1987, while Florida moved up from the nineteenth to the seventeenth wealthiest state in terms of per capita income. United States Bureau of the Census, *Statistical Abstract of the U.S.* (Washington, D.C.: Government Printing Office, 1989).

65. Interestingly, during our study in Ohio, Democratic Gov. Richard Celeste was in his second term (which expired in 1990), the gu-

bernatorial race featured a Democrat, Anthony Celebreeze, whose statewide reputation has been enhanced by his tough and nationally visible prosecutions of hazardous waste cases (see below). But Celebreeze comes from an old political family in Ohio, and his name is not as closely associated with hazardous waste regulation as Florio's.

66. After our study was completed, Florio became a controversial governor. His bold attempt to install a progressive income tax in New Jersey and an economy that slipped along with the rest of the country's fortunes as the 1990s began eroded his popularity, and he was defeated by Christine Whitman in the election of 1993.

67. The general "sensory prominence"—in terms of sight and smell—of New Jersey's petrochemical industry has always made it a source of both humor and vague concern. But concern was heightened by epidemiological studies labeling areas of the state a "cancer alley." Popular attention began to focus on the specific issue of hazardous waste in 1976—well in advance of national concern—when the state closed the Kin-Buc landfill in Edison Township in northeastern New Jersey. The site had been open for only three years when its seventy million gallons of wastes were found to be leaking into the groundwater.

68. These investigations revealed that the former owner of Chemical Control had been convicted of the "midnight dumping" of wastes from the site two years before the explosion and that, while under indictment, his firm had been bought out by reputed underworld interests. Newspaper and television stories also revealed that DEP had been aware of problems at Chemical Control since the early 1970s. An internal report of 1979 had warned that an explosion was likely. Indeed, a cleanup contract for the site had been signed in 1979. But the contract went to Coastal Services, a company that had not bid on the job and whose owner was related by marriage to the DEP official responsible for signing the agreement. There was evidence that the facility actually continued to accept hazardous wastes after DEP closed it for the cleanup! Fortunately, the limited cleanup that occurred before the explosion had at least removed the most dangerous substances found at the site.

69. Dan Grossman, "New Jersey Cleans Up Its Act," *Technology Review* 89 (April 1986), 11–13.

70. Indeed, the agency has gone through several additional reorganizations since we started visiting the state in 1983, most recently in 1986. In June 1989, Governor Kean named Christopher J. Daggett the new head of DEP; Daggett, most recently the administrator of the EPA's Region II office, which includes New Jersey, replaced Richard J. Dewling, who resigned on September 9, 1989.

71. The following illustrates our point about the renewed vigor of the state's redistributive efforts: a spokesperson for the governor commented that Kean did not see this five-year plan as part of a final solution and that "the governor would have liked to have seen this targeted more on the polluter, instead of public funds." Sure enough, just two years later, the legislature put together an additional $585-million cleanup package to supplement the 1986 funding scheme, once again balancing general revenue funding against industry taxation. *Environment Reporter* (October 31, 1986), 1033.

72. Of the $300-million bond issue, $100 million came from the original 1980 bond issue mentioned above. It had remained unspent because of legal challenges to its use brought by citizens' groups, which argued that the Spill Fund should be spent first. The 1986 referendum thus asked citizens to release the original amount for cleanup expenditures and approve an additional $200 million for the same purpose.

73. Northeast Hazardous Waste Project Report.

74. However, this analysis is tempered by the observation that, although many have tried, only one other state—Connecticut—has been able to pass similar legislation. Of course, land is an extremely valuable commodity in these two small, heavily populated, and generally wealthy states. Deborah L. Munt, "State-Initiated Hazardous Waste Management Programs: New Jersey's Environmental Cleanup Responsibility Act," *Innovations,* Published by the Council of State Governments, 1986.

75. A good example of this redistributive-allocational duality is the passage in 1986, be-

fore the approval of the five-year waste cleanup plan, of the state's Catastrophe Act. It immediately required two hundred industrial facilities to develop state-approved risk-management plans. The legislature appropriated $500 million from general revenues for initial implementation of the plan, but state agencies then began assessing and collecting fees from facility owners to reflect the operating costs of the program and establish it on an allegedly self-supporting basis. While the sustained entrepreneurial politics of hazardous waste regulation has made New Jersey a leader in innovative legislation, few other states have been able to follow its lead. Indeed, current economic recession and tax-cutting pressures in the state raise questions about New Jersey's continued ability to support this sort of regulatory effort. Further, although many states had tried, by the end of the decade only Connecticut had managed to pass an industrial cleanup bill like ECRA; neither the federal government nor any other state was able to pass legislation comparable to the Toxic Catastrophe Prevention Act. See Grossman, "New Jersey Cleans Up Its Act."

76. John J. Gargan, "Urban Revitalization and Vitalization: Testing Ohio's State and Local Government Capacity," in *Outlook on Ohio: Prospects and Priorities in Public Policy,* ed. W. O. Reichert and S. O. Ludd (Palisades Park, N.J.: Commonwealth Books, 1983).

77. Richard J. Aronson, *Financing State and Local Governments,* 4th ed. (Washington, D.C.: Brookings Institution, 1986).

78. The following passage from Carl Lieberman, "Ohio: The Environment for Political Activity," in *Government and Politics in Ohio,* ed. Carl Lieberman (Lanham, Md.: University Press of America, 1984), 7, summarizes this sentiment: "There is . . . an important moralistic subculture within the state, and the past tendency to accept reforms, particularly those which increase efficiency and reduce the costs of government, can be understood by the existence of this conception of politics."

79. Ohio had 32 sites, compared to Florida's 51 and New Jersey's 110 sites, as of June 1988.

80. Several sources we interviewed argued that such a revenue scheme created a conflict of interest for the OEPA. If it monitored HWDFs ag-

gressively and closed down those violating its regulations, OEPA would essentially be cutting off funding for its operations. There is no empirical evidence to indicate that such a conflict of interest actually occurred, but the credibility of the OEPA, already vulnerable, was further weakened by such speculation.

81. From our investigation into the compilation of these lists, however, it seems clear that the rankings of site severity were far from objective. Given the substantial demands for data about sites required for the EPA's hazard ranking system, sites without adequate data were simply ignored. The final rankings included only those sites for which state and local officials had been conscientiously collecting data. Indeed, out of 160 potentially dangerous sites identified by DER), only 27 were modeled by EPA. Despite the independent effect of missing data upon the ranking of Florida's sites, a limitation recognized initially by all who were familiar with the hazard ranking system, the final lists were treated in the policy process as if they reflected a complete and objective ranking. See Bruce A. Williams and Albert R. Matheny, "Hazardous Waste Policy in Florida: Is Regulation Possible?" in *The Politics of Hazardous Waste Management*, ed. Lester and Bowman.

82. As an example of its traditional sway in the legislature, wastes from phosphate mining were arbitrarily excluded from regulation under state hazardous waste laws.

83. The potential for such a marketing disaster was emphatically realized with the controversy over use of the pesticide Temik on the state's citrus in the 1980s. Actions by other states and nations to ban the import of Florida's citrus crop drove home to the industry the dangers flowing from publicity about contamination from pesticides and other hazardous substances.

84. In the fall of 1982, for example, Florida's speaker of the House of Representatives, Lee Moffitt (D-Tampa), appointed his close political ally and former state representative William Sadowski (D-Miami) to chair a task force on water quality. The task force issued its report in February 1983, and its findings garnered considerable media attention that spring, setting the stage for the consideration of an appropriate legislative response when the legislature convened in April of that year. The task force

report read like a blueprint for legislation, and that is precisely what Moffitt intended, for he had already identified the 1983 legislative session as the Year of Water Quality.

85. The *Time* cover story was particularly damaging. Entitled "Paradise Lost," it chronicled not only environmental problems, but ethnic strife and drug and crime problems as well. *Sports Illustrated*, in analyzing Florida's daunting water problems, suggested that drinking from a tap in the Sunshine State was like hooking up the kitchen faucet to the toilet.

86. The Florida House approved it by a vote of 116–4, and soon thereafter the Senate followed suit unanimously.

87. Florida had a much more difficult time than New Jersey in sustaining the momentum of entrepreneurial politics. As feared, hazardous waste did not remain at the top of the public agenda. After 1983, there were few dramatic, visible disasters to keep the issue in the spotlight, and other pressing public issues (e.g., drugs and crime) drew media attention in subsequent years.

88. The amendments lowered the small quantity exemption level from one thousand to one hundred kilograms per year for hazardous waste generators. This change hit Florida particularly hard because as much as 75 percent of all wastes in the state is generated by producers in the one-hundred- to one-thousand-kilogram range.

89. The starting salary for environmental specialists (e.g., biologists) in Florida was $1,261 per month, or $15,132 annually, in 1986, compared to $1,561 per month for the same position in Alabama and $1,510 per month in Georgia. For environmental engineers, the three-state figures were, $1,416, $1,995, and $1,776, respectively. Upon leaving DER, former employees generally found jobs paying 20 percent to 70 percent more, depending upon their areas of specialty (*Gainesville Sun* [May 21, 1986], 6A.)

90. For example, after garbage and medical waste floated onto New Jersey beaches in 1988, the agency was forced to divert time and money from its ongoing hazardous waste programs to establish an ocean waste program sponsored by the governor.

91. 475 U.S. 355 (1986).

92. The five companies, armed with the Su-

preme Court's ruling, went back to New Jersey and attempted to get back the money they had poured into the Spill Fund (an estimated $44,055,000), arguing that those funds had been spent to clean up sites eligible for Superfund moneys. The state Supreme Court held that the companies were not entitled to reimbursement but that the legislature had to reimburse the Spill Fund *from general revenues* for the above amount. This indirectly benefited industry because the taxing scheme of the Spill Fund would have doubled after the fund ran low. So the ruling shifted some of the burden from industry to the public, but the court's ruling left the structure of the Spill Fund intact. The U.S. Supreme Court declined to review this ruling.

93. Our discussion of right-to-know legislation in chapter 7 includes similar uses of the federal government's preemptive powers.

6 / Siting of Hazardous Waste Disposal Facilities

1. Legislative Commission on Toxic Substances and Hazardous Wastes, *Hazardous Waste Facility Siting* (Albany: New York State Legislature, 1987), 25.

2. This typology is a simplified version of the classification schemes of Susan Hadden et al., "State Roles in Siting Hazardous Waste Disposal Facilities: From State Preemption to Local Veto," in *The Politics of Hazardous Waste Management*, ed. James P. Lester and Ann O'M. Bowman (Durham: Duke University Press, 1983); Richard N. L. Andrews, "Hazardous Waste Facility Siting: State Approaches," in *Dimensions of Hazardous Waste Policy*, ed. Charles E. Davis and James P. Lester (New York: Greenwood Press, 1988).

3. Florida's siting strategy essentially tried to "learn" from the failures of the comprehensive and case-by-case approaches, and so is difficult to characterize in terms of our typology. The Florida case is peculiarly instructive in its successes and failures, and these will be dealt with at length in the study of Florida's siting program, below.

4. The specter of environmental injustice, sometimes referred to as environmental racism, is perhaps the most potent and well-documented objection to past HWDF siting, and

the charge emerges precisely because the pluralist conception of fair process ignores the high cost of educating and organizing poor (and often minority) communities to the point where they can participate equally, at the level assumed for effective participation in pluralist terms. See generally Robert D. Bullard, *Dumping in Dixie: Race, Class and Environmental Quality* (Boulder: Westview Press, 1990); Robert D. Bullard, ed., *Unequal Protection: Environmental Justice and Communities of Color* (San Francisco: Sierra Club Books, 1994).

5. Michael L. Poirier Elliott, "Improving Community Acceptance of Hazardous Waste Facilities through Alternative Systems for Mitigating and Managing Risk," *Hazardous Waste* 1 (1984), 397–410.

6. Elliott's summary of the third proposal (from "EMI") is quoted entirely here: "EMI argued that the risks of hazardous waste treatment developed not so much because of inadequate technology, but because of less-than-ideal management practices. They offered to open the operations of the company to public scrutiny and to subject safety decisions to community review. The core of their proposal was a safety board on which community residents would sit. The board would oversee the safety of the plant, manage its own annual budget for making improvements, and have emergency powers should hazards develop. The facility and its records would be inspected by an engineer hired by the town. Payments would be made to the town fire department so that it would have the specialized equipment and training to cope with emergencies. Agreements on how to resolve disputes would also stipulate the creation of emergency action trust funds to ensure the availability of necessary funds. Finally, EMI indicated that it would own all delivery trucks, specifying the routes they could travel and the hours they could operate. By carefully attending to issues of liability, accessibility and open management, EMI offered reassurances that no shortcuts would be taken that might undermine the safety of the plant." Ibid., 399–400.

7. Ibid., 408.

8. Substantive participation in HWDF management might not actually reduce the risks faced by the surrounding community, but that community would at least have shifted its perception of the risks from an "involuntary" to a

"voluntary" consciousness. This psychological shift is crucial for overcoming citizen distrust, given that the public is apparently more tolerant of voluntary risks than they are of equal involuntary risks. See Baruch Fischoff et al., *Acceptable Risk* (New York: Cambridge University Press, 1981).

9. On this point, see Michael R. Greenberg and Richard F. Anderson, *Hazardous Waste Sites: The Credibility Gap* (New Brunswick, N.J.: Center for Urban Policy Research, 1984).

10. Our discussion of early state efforts in this section draws upon our own research and the early study of siting in New Jersey by David Morell and Christopher Magorian, *Siting Hazardous Waste Facilities: Local Opposition and the Myth of Preemption* (Cambridge: Ballinger Publishers, 1982).

11. Ibid., 28–38.

12. Significantly, the successful siting in Bridgeport occurred before the Love Canal incident, while the failed effort in Bordentown came in its immediate aftermath. But New Jersey's citizens were well aware of the dangers associated with hazardous waste disposal long before the rest of the nation, so we feel that the lessons learned from these examples remain relevant to our analysis.

13. Although the Monsanto siting occurred before the nationally publicized events at Love Canal, New Jersey citizens were already aware of the risks of hazardous waste disposal as a result of the highly publicized controversy over the Kin-Buc landfill, discussed above.

14. Contrast these first two steps with the stated rationale of the HWFAB in Ohio, discussed in our next case study, where the point of the process was to avoid publicity and move things along quickly before local opposition could develop.

15. Once again, the contrast with the CECOS / CER site in the Ohio case study is interesting. There, the conflict between the locality and the site operators intensified as ownership passed from a local firm to unfamiliar corporations, increasingly distant from the community.

16. SCA ran a controversial solid waste landfill in Bordentown, and, as a result, the corporation was perceived as remote and unresponsive to local concerns.

17. Public suspicion about the safety of hazardous waste disposal came to a head in 1980, when East Windsor was identified as a possible site for an HWDF. Local opposition rapidly developed, and the largest public meeting in the township's history abruptly ended the proposal. Soon thereafter, a similar local protest blocked a possible site in Alloway Township. See Morell and Magorian, *Siting Facilities*, 24.

18. Cited in ibid., 162.

19. Whereas the northeastern portion of the state is highly industrialized, much of the south and extreme northwestern portions of the state is still farmland or wealthy suburban enclaves. The powerful statewide environmental organizations had a difficult time reconciling the conflicting pressures brought to bear by residents in these very different regions. In particular, they were perceived by many local activists in the more urbanized areas as being biased toward the interests of the wealthier suburban regions of the state.

20. Although the commission's function is similar to that of the Ohio HWFAB, discussed below, the logic of its membership is quite different and illustrates the difference between pluralist and managerial perspectives. Consistent with managerial logic, the Ohio board is composed only of representatives of state agencies and technical experts.

21. Cited in Morell and Magorian, *Siting Facilities*, 237. This standard allowed the court much greater discretion than was the case in Ohio's siting process, discussed in our next case study.

22. "This site designation process . . . is essentially preemptive; although local parties can provide input, at no point can the community render a decision of its own on the proposed site designation." Ibid.

23. Ibid., 112.

24. Patrick G. Marshall, "Not In My Back Yard!" *Editorial Research Report* (Washington, D.C.: Congressional Quarterly, 1989), 314.

25. James F. McAvoy, "Hazardous Waste Management in Ohio: The Problem of Siting," *Capital University Law Review* 9 (1980), 447–48. Echoing the thoughts of regulators from as far back as Charles Francis Adams, McAvoy argues that participation by interested groups would thwart good policymaking by experts.

26. To avoid the appearance of being a rubber stamp and as evidence of the symbolic importance of language, the name of this board

was later changed to the Hazardous Waste Facilities Board, i.e., the word "Approval" was dropped.

27. The legislation strictly prohibited localities from interfering with the HWDF siting. It banned local attempts to adopt exclusionary zoning policies or to impose requirements upon the HWDF beyond those defined by the state.

28. The court could hear additional evidence on appeal to supplement the record only if such evidence could not have been known during the hearing.

29. This is a relatively lenient standard of proof, generally meaning a preponderance (51 percent) of the evidence supporting a reasonable, *but not necessarily the best*, conclusion, given the facts in the record. A stronger standard, for example, would have been "clear and convincing evidence," implying a higher standard of proof, somewhere between "substantial evidence" and evidence "beyond a reasonable doubt."

30. Note that granting a compliance permit to an existing facility is quite different from siting a new facility. Ohio was no more successful at the latter than any other state, but much more successful at the former.

31. Our discussion of these events draws from the summary in Michael E. Kraft and Ruth Kraut, "Citizen Participation and Hazardous Waste Policy Implementation," in *Dimensions of Hazardous Waste Politics and Policy*, ed. Charles E. Davis and James P. Lester (New York: Greenwood Press, 1988), as well as our own research.

32. The site was later purchased by Chemical and Environmental Conservation Systems (CECOS), hence the reference to CECOS/CER in our discussions of the site itself as distinct from its original developer, CER.

33. Kraft and Kraut, "Citizen Participation," 69.

34. One of the disadvantages of OEPA moving so swiftly to approve the facility was that Ohio rapidly became a net importer of hazardous wastes, as other states saw the Ohio dump as an attractive alternative to building or permitting their own facilities. The incentive for states to "beggar-thy-neighbor" is one persuasive argument for a stronger federal role in managing hazardous waste disposal. In addition to making the site controversial among citizens' groups in

Ohio (because it accepted wastes from other states), it was the inability of other states to build such facilities that gives the Ohio case its national implications.

35. Although it was true at the time that I-CARE did not have access to technical expertise or information and that their concerns were based mainly on fear and suspicion, little energy was spent by state officials or facility operators either to inform the group or to address its concerns seriously.

36. This response contrasts sharply with the negotiating strategy employed by Monsanto in the Bridgeport, New Jersey, case described above. In Ohio, officials incorrectly assumed that having access to adequate information automatically leads to public agreement over policy. They thought the reason that citizens raised objections was simply because they had no access to such information (and would not understand it if they did). The Ohio response, while not surprising, is based on flawed assumptions about the role of technical information in the policy process. Officials assumed that issues such as operating HWDFs safely are strictly technical in nature and that mastery of that technical body of knowledge is required before any public input can be taken seriously, even when fundamental questions are raised about the health and safety of communities.

37. Kraft and Kraut summarize the situation: "The underlying issues were ignored as the antagonists resorted to legal maneuvers and largely symbolic posturing. Such actions did little to foster public understanding of technical and administrative concerns or to create more constructive processes for citizen participation." Kraft and Kraut, "Citizen Participation," 70. We consider "strategic lawsuits against public participation" (SLAPP suits) as an anticommunitarian strategy in chapter 7.

38. I-CARE's alliance strategy is typical of the larger grassroots movement against toxics, a subject dealt with in detail in chapter 7.

39. The conflict had national implications not only because the site was owned by a multinational corporation, but also because, as one of a small number of licensed and privately owned landfills, CECOS/CER was crucial to early national approaches to the hazardous waste problem. Officials in New Jersey mentioned the CECOS/CER site as the most desirable place to

ship hazardous wastes and, with the threatened closing of sites in Alabama and South Carolina to Florida's waste exporters, Florida officials also mentioned the Ohio site as the nearest possible alternative.

40. Emblematic of the degree to which control of the site was moving away from the community level, this expansion, following a series of approvals by the EPA from 1978 to 1981, was paid for by a grant from the Ohio-Kentucky-Indiana Regional Council of Governments.

41. Phenol was found in the water pumped out of the CECOS / CER cell and in the town's water tanks, although none was detected in the intervening stream.

42. The prosecution of the BFI criminal case was a complicated and ultimately frustrating affair that ended in a mistrial on ninety-six criminal charges in 1988. Remarkably, while under indictment, BFI attempted to renew CECOS/CER's RCRA permit. OEPA denied the request, referring to the landfill's "past compliance history," among other things. One effect of this series of events was that citizens and environmental groups began to express some confidence in OEPA and the state generally (particularly after the attorney general's action) on the issue of hazardous waste regulation.

43. Ironically, at the end of the decade, CECOS / CER continued to accept EPA-permitted PCB wastes because they are under the jurisdiction of TSCA, not RCRA.

44. As was the case in New Jersey, this provision reflected the dissatisfaction of local activists with the existing process's pluralist approach to representation on the HWFAB, whereby only local elected officials were appointed when a site was designated in their jurisdiction.

45. Ironically, preemption was supported by many local elected officials who feared being unable to make responsible siting decisions because of the likelihood of being held politically accountable by outraged local citizens. This echoes the sentiments of many local officials in both Florida (e.g., Union County) and New Jersey (e.g., Newark) who were happy to get the economic payoffs for siting a facility in their community, as long as they could not be held accountable for issues of safety. Local governments operate most comfortably when responding to distributive, or allocational, is-

sues like economic development. They find it difficult to accommodate citizen reactions to redistributive policies and thus are relieved to shift the "blame" to another level of government, while yet receiving whatever "benefits" accrue from those policies.

46. As a sop to frustrated local interests, the legislation gave OEPA the authority to ban landfilling in Ohio by 1987, but only if it could show that "technically feasible and affordable alternatives" were available. The 1984 legislative session also produced authorization and funding for hiring full-time OEPA personnel to monitor all HWDFs in Ohio; previously, such monitoring had occurred on only a limited basis (as at CECOS / CER). The value of this legislation was demonstrated only two months later, when the OEPA employee at CECOS / CER discovered the pumping incident, discussed above.

47. The constant changing of informational requirements made the entire process of litigating siting decisions quite expensive. By 1991, the city of Dayton had spent more than $100,000 appealing a board decision.

48. Poor work by the board in developing the permit for the proposed facility in Dayton led the EPA to remand it for reformulation, costing Dayton and the state a considerable amount of money for new studies. The city of Toledo experienced similar problems.

49. Moreover, more sophisticated monitoring and management experience at sites like CECOS / CER gave environmentalists new credibility in the policy process, at the expense of OEPA's and the HWFB's credibility, particularly on the issue of landfilling.

50. Although authorized, these funding and staffing increases were long delayed, as noted in chapter 5.

51. Other Amnesty Days collections have been carried out since, and many counties have continued similar programs in the 1990s, after the state lost interest with the collapse of the state HWDF siting process.

52. Again, we should emphasize that the program's success has been confined to raising public awareness. In terms of addressing the actual hazardous waste problem in Florida, about one-tenth of 1 percent of the state's total population deposited during Amnesty Days roughly one-tenth of 1 percent of all the hazardous wastes generated annually in the state.

53. We return to these incentives and their connection to the strategies of grassroots groups protesting HWDFs in chapter 7.

54. It was later disclosed that in December 1987, the mayor had bought land adjacent to the site and convinced the city commissioners to consider annexing the site in order to have greater control over the site and to take advantage of the 3 percent set-aside (for host communities) from the HWDF's revenues, as provided under the WQAA.

55. On January 26, 1988, three weeks after the five sites were nominated, DER held the only local meeting about the HWDF. The nominations were presented as a fait accompli, and DER simply explained the site selection process and discussed the operation of the proposed facility with angry residents.

56. The contrast between that county's official reluctance to challenge the site at the hearing and the fast-growing grassroots opposition to the site within its borders illustrates perfectly how local governments are often completely unprepared to absorb community reaction to redistributive decisions. As was the case in New Jersey and Ohio, it also demonstrates the degree to which elected officials are quickly perceived by citizen groups as illegitimate representatives of local interests. Union County, with an economy based largely upon state prison operations, had a history of accepting projects unwanted by other parts of the state, traditionally viewing them as opportunities for economic growth. But residents drew the line at accepting the state HWDF. The county had passed a resolution against the plant over a month earlier; yet county officials' comments (about not attending the Jacksonville hearing) indicated that they had been distracted by turf battles with Raiford, the community closest to the site, over which jurisdiction should annex the site, in order to benefit from the revenues of the HWDF's operation.

57. DER repeatedly asserted confidence in the Westin, Inc., study, but records showed great dissatisfaction within DER over the consultant's performance, particularly with the quality of the consultant's employees, their preparation for the job, and their inability to do the work promised in a timely fashion. After DER announced the five candidate sites in January, it could not respond to citizens' objections about

its choices because Westin, Inc., had not sent any of the material supporting the sites selected to DER. The supporting material did not arrive until after the January 26 meeting between DER and local residents, mentioned above. In a remarkable memo regarding the consultant's unpreparedness, the DER complained that when Westin, Inc., officials arrived in north central Florida on January 20, 1988, to investigate the already nominated sites, "they came without any maps of the sites or information and documentation. Richard Deadman [a DER representative] had to instruct them to go buy U.S. Geological Survey quad sheets and maps [to locate the sites]" (quoted in the *Gainesville Sun,* April, 27, 1988), 14A.

58. *Gainesville Sun,* March 18, 1988, 6A.

59. Ibid., March 17, 1988, 5B.

60. Ibid., March, 18, 1989, 6A.

61. Such an approach to regulatory policymaking, of course, echoes the arguments made by the prophets of regulation—Adams, Landis, and Kahn—outlined in chapter 2.

62. *Gainesville Sun,* May 19, 1989, 1A.

63. Ibid., June 25, 1989, 1G.

64. The Union County location, while remote from sources of hazardous waste generation in Florida, is close to the Georgia border. At about the same time the legislature ratified the Union County site, DER entered into unpublicized negotiations with the EPA and other southeastern states to develop an interstate agreement for handling, treating, and disposing of hazardous waste, meaning that Florida would be processing not only its own waste at the as-yet-unconstructed Union County HWDF, but also the wastes of other states. During the entire siting process, DER had emphasized only that Florida should be responsible for handling its own wastes. The issue of wastes coming into the Union County HWDF from out of state was not raised during the site selection process, and it was not discussed in the legislature. The proximity of the Union County site to the Georgia border makes interstate use much more feasible than would a location nearer to most of Florida's hazardous waste generators in the southwestern and southeastern parts of the state. Although this point was never mentioned during the site selection process, the Union County location would probably be economically more attractive

to prospective private operators of the facility than would a south Florida location, precisely because it would encourage out-of-state users to ship wastes into Florida, as well as offering the only source for the disposal of hazardous waste from within the state. From our earliest interviews with DER officials and legislative staffpeople in Florida, their constant concern was to make HWDF options attractive enough to encourage private companies to operate such facilities.

65. Here, the ways in which the outcome of a flawed process can become accepted as correct if that process is labeled technical and objective is reminiscent of the uses of the rankings of Florida's sites for the Superfund interim NPL (see discussion in chapter 5).

66. *Gainesville Sun,* June 25, 1989, 4G. Twachtmann also pointed out that the legislature's streamlining of the HWDF appeal process applies to all HWDF facility appeals, not just the appeal of the Union County site. Thus, opponents of future sites would have their objections compressed into one final hearing, a decidedly difficult mechanism for the delicate and time-consuming process of developing a dialogue—much less a consensus—among contending parties.

67. *Gainesville Sun,* May 19, 1989, 5A.

68. Ibid.

69. Ibid., 1A.

70. Ibid.

71. All quotes from *Gainesville Sun,* June 1, 1989, 1A.

72. In a classic example of federalism at work, key legislators excused their preemptive behavior in the Union County affair by claiming that the site was necessary to ensure that federal Superfund moneys would not be lost. In fact, the money was never cut off; but the legislators were afraid of jeopardizing an external revenue source at a time when "No new taxes" was a persistent campaign cry heard around an increasingly Republican state.

7 / Not-in-My-Back-Yard, Right to Know, and Grassroots Mobilization

1. Perhaps the most accessible conventional journalistic treatment of the NIMBY issue is Charles Piller, *The Fail-Safe Society: Community Defiance and the End of American*

Technological Optimism (New York: Basic Books, 1991). See also Patrick G. Marshall, "Not In My Back Yard!," *Editorial Research Report* 1, no. 21 (Washington, D.C.: Congressional Quarterly, 1989), 305–19, for a brief summary of conventional criticism of NIMBY as citizen protest. For a more intensive and empirically sophisticated scholarly analysis, see Kent E. Portney, *Siting Hazardous Waste Treatment Facilities: The NIMBY Syndrome* (Westport, Conn.: Auburn House/Greenwood, 1991).

2. On the relative ranking of environmental risks by laypeople versus experts, see EPA administrator William Reilly's remarks in William K. Stevens, "What Really Threatens the Environment," *New York Times,* January 29, 1991, C 4; see also Peter Passell, "Are Waste Cleanups Worth Cost?" *Gainesville Sun,* September 1, 1991, 1A. A charitable and insightful treatment of NIMBY and the failure of HWDF siting that nevertheless originates in the managerial mode is Michael O'Hare, Lawrence Bacow, and Debra Sanderson, *Facility Siting and Public Opposition* (New York: Van Nostrand Reinhold, 1983). The authors conclude with the recommendation that successful siting might occur only if government and developers have negotiated adequate economic compensation for the affected community. The premises for their analysis are consistently managerial: "Our view of the facilities siting problem is characterized by two basic propositions: 1. Inadequate mechanisms exist at present for the parties affected by a new facility proposal to share in the benefits the project will provide to society as a whole, or to effectively negotiate the size of their share [faulty risk comparisons]. 2. Much of the facility siting debate is ignorant or ill-informed because the social, political, and economic structures by which information is made available obstruct its efficient use or generate [inadequate or misunderstood information]." Ibid., 3.

3. Marshall, "Not In My Back Yard!" quoting Douglas Porter of the Urban Land Institute,: "I think there is a role for citizen protest. And there are extraordinary times when it ought to be exercised. But to achieve the kind of constant use that NIMBY has achieved seems to go right around the whole representative government process" (307). See also Diane Graves's remarks criticizing the local

resistance to New Jersey's siting process and Florida Gov. Bob Martinez's statement regarding the Union County site mentioned in chapter 6.

4. For an insightful critique of the seemingly neutral suggestion of regulatory negotiation, see Christine B. Harrington, "Regulatory Reform: Creating Gaps and Making Markets," *Law and Policy* 10 (1988), 293–316.

5. Public involvement is especially striking in an era of increasing public disaffection with the conventional forms of political participation. See Benjamin Ginsberg and Martin Schefter, *Politics by Other Means* (New York: Basic Books, 1990).

6. Our discussion in chapter 2 introduced this point, referring to the thought of James Wilson, discussed in Jennifer Nedelsky, *Private Property and the Limits of American Constitutionalism: The Madisonian Framework and Its Legacy* (Chicago: University of Chicago Press, 1990), and the history of grassroots reform efforts in American politics developed in James A. Morone, *The Democratic Wish: Popular Participation and the Limits of American Government* (New York: Basic Books, 1990). Robert Bellah et al., *Habits of the Heart: Individualism and Commitment in American Society* (Berkeley: University of California Press, 1985), also clarify the communitarian perspective by emphasizing the communitarian understanding of government's role as a preserver of public values over the more individualistic conception of government as a promoter of private rights. This important distinction becomes particularly relevant as we address public versus private control of information in the policy process below.

7. Previous chapters demonstrate that the consequences of social regulation are likely to transgress unrecognized barriers of market or government operation. The recognition of externalities, the ambiguity of risks, and the redistribution of consequences all strain the conventional relations among market, government, and technology. Especially in the area of hazardous waste regulation, where local populations are asked by government to accept the burdens of dangerous market practices, public challenges should be expected. And their success is more likely because, as we explain in chapter 4, they are reinforced by the federal

system's structural reluctance to implement redistributive policies at the local level. This point involves some irony, and we return to it later in this chapter.

8. Morone, *The Democratic Wish*, generally.

9. This view of information and its use by citizens is strikingly similar to John Dewey's approach, discussed in chapter 3. Recall that Dewey saw science not simply as a body of knowledge used by elites and produced outside the democratic process, but as a technique for critical reasoning that needed to be taught to citizens in a democracy. Dewey also saw the connection between the structure of the workplace and the possibilities for meaningful democratic participation—hence his calls for the reform of industrial practices.

10. Consistent with the communitarian thrust of this argument are some of Piller's conclusions in *The Fail-Safe Society*, 195: "Society would do well to stop trying to 'solve the Nimby problem' and to begin exploring ways to make Nimbyism unnecessary. This will require a radical reappraisal of the relationship between the public and the entire scientific and technological enterprise.

Nimbyism tests the burden of proof regarding scientific and technological risk. Formerly, victims were compelled to prove harm. Increasingly, scientists and technocrats take the defensive in efforts to prove safety. But this effect has been limited, haphazard, and often disruptive without social gain. To prepare the ground for a consensus on risks and benefits— one based on trust and shared power—the burden of proof must be shifted deliberately, methodically, and definitively."

11. Quoting George White, vice president for public affairs at Phoenix Advertising in Durham, North Carolina. His firm was assisting ThermalKEM, Inc., of Rock Hill, South Carolina, in its failed attempt to site a hazardous waste incinerator in Pender County, North Carolina. Reported in Sean Loughlin, "Waste Triggers Red Flag," *Gainesville Sun*, November 18, 1991, 1A, 8A.

12. In *Euclid v. Ambler Realty Co.*, 272 U.S. 365 (1926), the U.S. Supreme Court first recognized that communities could control land uses politically through zoning without violating constitutional property rights.

13. The consequences of suburban develop-

ment to the political and social ordering of urban areas are exposed in Constance Perin, *Everything in Its Place: Social Order and Land Use in America* (Princeton: Princeton University Press, 1977).

14. The sometimes extraordinary efforts of suburban homeowners to resist subsidized housing development are discussed in Neil K. Komesar, "Housing, Zoning, and the Public Interest," in *Public Interest Law: An Economic and Institutional Analysis*, ed. Burton A. Weisbrod (Berkeley: University of California Press, 1978), 218–50.

15. "Not In My Back Yard: Removing Barriers to Affordable Housing," report to President Bush and Secretary Kemp by the Advisory Commission on Regulatory Barriers to Affordable Housing (Washington, D.C.: Government Printing Office, 1991).

16. Ibid., 1–5.

17. Portney, *Siting Hazardous Waste Treatment Facilities*, 71–75, tables 4.1–4.3; 84–85, table 4.9. His analysis is based upon surveys conducted in the early 1980s of respondents from five Massachusetts cities that had not experienced any particular exposure to hazardous substances. Earlier (11), Portney defines "the NIMBY Syndrome" as "a reflection of a public attitude that seems to be almost self-contradictory—that people feel it is desirable to site a particular type of facility somewhere as long as it is not where they personally live."

18. See, for example, Norman R. Luttbeg and Michael D. Martinez, "Demographic Differences in Opinion, 1956–1984," in Samuel Long, ed., *Research in Micropolitics: A Research Annual* 3 (1990), 83–117, esp. 102–04, and table 5.

19. Portney, *Siting Hazardous Waste Treatment Facilities*, 91–92, table 4.14.

20. Ibid., 85–88, tables 4.10–4.11.

21. In Portney's Massachusetts survey, 62 percent of those opposed to HWDF siting were opposed to siting anywhere in the state, compared to 38 percent opposed only to siting in their communities. The figures for his national survey were 58 percent and 42 percent, respectively.

22. Janice E. Perlman, "'Grassroots Participation from Neighborhood to Nation," in *Citizen Participation in America*, ed. Stuart Langton (Lexington, Mass.: Lexington Books,

1978), 67. Perlman warns that citizen action groups are different from citizen involvement groups—a mainstay of pluralist legitimacy—in that the latter are organized from the top down: "Individual citizens . . . are often asked [under a citizen involvement strategy] to participate at considerable personal sacrifice in public hearings or on local boards, only to find themselves as powerless as before. The citizen-action approach, then, is based on a substitution of numbers for monetary resources and of commitment and courage for position and authority" (66).

23. Ibid., 69–72, esp. table 6–1. Perlman's sympathetic treatment of citizen action groups reflects earlier research on citizen participation that reaches conclusions consistent with some of our requirements for the dialogic model of policy-making discussed in chapter 3: "In most cases where power has come to be shared it was *taken* by the citizens, not *given*. . . . Partnership can work most effectively when there is an organized power-base in the community to which the citizen leaders are accountable; when the citizen group has the financial resources to pay its leaders reasonable honoraria for their time-consuming efforts, and when the group has the resources to hire (and fire) its own technicians, lawyers and community organizers." Ibid., 67, quoting Sherry Arnstein, "A Ladder of Citizen Participation," *Journal of the American Institute of Planners* 35 (July 1969), 221 (emphasis in original).

24. See Susan G. Hadden, *A Citizen's Right to Know: Risk Communication and Public Policy* (Boulder: Westview Press, 1989), 5–11, for a succinct discussion of the relation between toxics and cancer.

25. The Delaney Clause in the 1958 Food Additives Amendment to the Act provides that "no additive shall be deemed to be safe if it is found to induce cancer when ingested by man or animal, or if it is found, after tests which are appropriate for the evaluation of the safety of food additives, to induce cancer in man or animal" [21 U.S.C. 348(c)(3)(A)(1976)]. For a discussion of the controversy surrounding the Delaney Amendment, see generally Thomas O. McGarity, "Substantive and Procedural Discretion in Administrative Resolution of Science Policy Questions: Regulating Carcinogens in EPA and OSHA," *Georgetown Law Journal* 67 (1979), 729–810.

26. Rachel Carson, *Silent Spring* (Boston: Houghton-Mifflin, 1962). Of course, other events during this period dramatized the effects of toxins on human health as well as on the environment. In 1956, there was broad media coverage of the severe neurological damage to those who had eaten fish contaminated with high levels of mercury in Minamata Bay caused by industrial water pollution in Kyushu, Japan. Again, in the early 1960s, media coverage of so-called Thalidomide babies graphically illustrated the dangers of improperly tested pharmaceuticals on fetuses.

27. As federal air and water pollution controls were put in place, the pollutants captured became hazardous waste, which initially could be disposed of in a virtually unregulated fashion.

28. Evidence came largely from laboratory studies conducted on small mammals and broader epidemiological studies of cancer rates among human subpopulations. Compare two influential occupational disease policy studies done during the 1970s: Nicholas A. Ashford, *Crisis in the Workplace: Occupational Disease and Injury* (Cambridge: MIT Press, 1976), and David Doniger, *The Law and Policy of Toxic Substances Control: A Case Study of Vinyl Chloride* (Baltimore: Johns Hopkins University Press, 1978).

29. The 1976 toxics laws were something of a congressional afterthought, and the lack of urgency surrounding their passage reflected in part congressional disenchantment with the traditional leadership role of the federal government as a pioneer in a new area of social regulation. See chapter 5, generally, and Samuel S. Epstein, Lester O. Brown, and Carl Pope, *Hazardous Waste in America* (San Francisco: Sierra Club Books, 1982), 189–94.

30. See David P. McCaffrey, OSHA and the Politics of Health Regulation (New York: Plenum Press, 1982), and Charles Noble, *Liberalism at Work: The Rise and Fall of* OSHA (Philadelphia: Temple University Press, 1986), for discussions of OSHA's beleaguered attempts to regulate toxics in the workplace.

31. For a thorough analysis of the Interagency Regulatory Liaison Group, see Marc K. Landy, Marc J. Roberts, and Stephen R. Thomas, *The Environmental Protection Agency: Asking the Wrong Questions* (New York: Oxford

University Press, 1990), 173–203. Mark E. Rushefsky, *Making Cancer Policy* (Albany: SUNY Press, 1986), focuses on the political evolution of the Cancer Policy itself from its inception to its effective demise in the Reagan administration. See also Albert R. Matheny and Bruce E. Williams, "Regulation, Risk Assessment, and the Supreme Court: The Case of OSHA's Cancer Policy," *Law and Policy* 6 (October 1984), 425–49.

32. Walter A. Rosenbaum, *Environmental Politics and Policy* (Washington: Congressional Quarterly Press, 1985), 64–70, esp. 66–67, table 2–3: "Public Concern with Environmental and Other Domestic Issues, 1972–1980."

33. Typical of this sort of criticism is William Tucker, *Progress and Privilege: America in the Age of Environmentalism* (Garden City, N.J.: Anchor Press/Doubleday, 1982).

34. A representative treatment of toxics issues by a print journalist is Michael Brown, *Laying Waste: The Poisoning of America by Toxic Chemicals* (New York: Pantheon, 1980). Landy, Roberts, and Thomas provide a critical portrait of television news coverage of hazardous substance issues in *The Environmental Protection Agency*, 23–24.

35. For the importance of NIOSH as a basis for workers' information, see our discussion in chapters 2 and 3. The NIOSH Registry of Toxic Substances covers well over 50,000 substances encountered in the workplace. Under the National Occupational Hazards Survey published in 1976, NIOSH listed 64,891 "exposure agents." Only half of these contained chemicals actively regulated by OSHA.

36. Covering a conference on COSH groups in 1982, Gail Robinson, "COSH!" *Exposure* 21/22 (August/September 1982), 2, noted other disagreements among these groups as well: "COSH members, for example, disagree over the role of unions in the movement. Some feel that unions must be the constituency while others believe that COSH groups should actively encourage the participation of community groups, minority organizations and others." See also, Noble, *Liberalism at Work*, 134.

37. Noble, *Liberalism at Work*, 134, comments, "Many COSH members were New Leftists, and they brought a more radical vision to the problem of working conditions than most union officials did. The COSHs argued for

worker control over working conditions and encouraged rank-and-file organization and participation as they lobbied for legislative reforms and educated union locals about health hazards."

38. Labor often pursues redistributive issues at the national level, and certainly the OSHACT is an example of just such an effort, engaging precisely the question of control of the workplace, but traditionally the national political agenda of unions has been consistently absent from the local labor agenda: see J. David Greenstone, *Labor in American Politics* (New York: Knopf, 1969), 170–75. Daniel Berman, *Death on the Job: Occupational Health and Safety Struggles in the United States* (New York: Monthly Review Press, 1981), found in the 1970s that while rank-and-file union members consistently ranked job safety as a priority concern, union leaders did not.

39. Susan G. Hadden, *A Citizen's Right to Know*, 24. See also note 44 below. A key role in the Philadelphia meeting was played by the Delaware Valley Toxics Coalition, an early and influential grassroots toxics group bringing together skilled community and labor activists with scientific experts on hazardous substances.

40. As a form of regulation, requiring industry to provide product information is as old as the Federal Food and Drug Act of 1906 and hardly seems revolutionary. Labeling regulation, stretching from the list of ingredients in processed foods to clarity in loan agreements and to product safety labeling, may all be considered a part of traditional economic regulation insofar as its aim is to correct for a "market failure" by meeting the "free market" assumption of informed purchase in private exchanges between a producer and consumer. This may be contrasted with right to know involving access to information for public and political decision making. For examples of the latter, see the next note.

41. David M. O'Brian, *The Public's Right to Know: The Supreme Court and the First Amendment* (New York: Praeger, 1981), 38, cited in Hadden, *A Citizen's Right to Know* , 4, fn. 1. Right to know, in this sense, is more accurately represented in the Federal Freedom of Information Act of 1966 and the Government Sunshine Act of 1976 than in worker and community right to know.

42. Nedelsky, *Private Property,* uses Wilson's vision of politics as a basis for advocating a new vision of the relation between property and polity. She argues that "the possibility of disassociating the founding connection between property and limited government also invites a new way of conceiving and implementing the competing values of democracy and individual rights, of collectivity and individuality, so that both are fostered" (272).

43. Employee right to know would require industry to provide workers with information about the toxics they confronted in the workplace, and community right to know would give citizens and local governments (particularly emergency response personnel) access to information about the use, storage, and transportation of toxics within their jurisdictions.

44. Ken Silver, "The Right-to-Know Story: No Easy Victories," *Exposure* 28/29 (March/April 1983), 4. He also describes the effort: "In workplaces, union halls, shopping malls, bowling alleys, laundromats and other places where working people gather, COSH activists in Philadelphia, Chicago, Massachusetts and California gathered tens of thousands of signatures on a national right-to-know petition. In doing so, they injected the term 'right-to-know' into trade union parlance" (4).

45. Hadden, *A Citizen's Right to Know,* 22–25. The states with some form of right-to-know legislation in 1984 were Alaska, California, Connecticut, Delaware, Florida, Illinois, Maine, Maryland, Massachusetts, Michigan, Minnesota, New Hampshire, New Jersey, New York, Pennsylvania, Rhode Island, Virginia, Washington, West Virginia, and Wisconsin. Hadden argues that the spread of right-to-know agitation followed local accidents or emergencies dramatizing the presence of hazardous substances. But COSHs were equally instrumental in the rapid spread of this type of legislation, virtually all of which passed before the dramatic Bhopal, India, accident in 1984, which inspired passage of federal right-to-know legislation. See, e.g., Gail Robinson, "COSH!" *Exposure* 21/22 (August/September 1982), 1–3; Ken Silver, *Exposure* 28/29 (March/April 1983), 6–7; and David H. Wegman, "COSH: A Grass-Roots Public Health Movement," *American Journal of Public Health* 74 (1984), 964–65.

46. Michael S. Brown, "Disputed Knowl-

edge: Worker Access to Hazard Information," in *The Language of Risk: Conflicting Perspectives on Occupational Health*, ed. Dorothy Nelkin (Beverly Hills, Calif.: Sage, 1985), makes the point straightforwardly: "The organizational structure of firms . . . limits promotion of health and safety. Managers with the authority to make investment decisions are quite removed from the day-to-day operations. Line supervisors who are familiar with conditions are responsible for production and evaluated on their ability to maintain productivity. Communicating information about hazards may lead to worker concerns and complaints, thereby reducing productivity. There is little incentive to openly discuss health and safety problems."

47. Noble, *Liberalism at Work*, 25.

48. For example, see Bob Sanders, "Corporate Coalition Challenges New Jersey Law," *Exposure* 42 (November/December 1984), 7–8, 10, which discusses industry lobbying in several states with pending right-to-know legislation. In chapter 6 we noted this occurrence in New Jersey, where industry lobbyists were so distracted by fighting passage of the New Jersey right-to-know law that they largely neglected the ECRA bill during that session, which turned out to be a much more troublesome and costly piece of legislation for industry.

49. Key allies in the right-to-know movement at the local level have been public safety officials (particularly fire departments) concerned about the lethal prospects of industrial accidents. The more general success of the new coalition between workers and residents is extensively discussed and documented in Jeremy Brecher and Tim Costello, eds., *Building Bridges: The Emerging Grassroots Coalition of Labor and Community* (New York: Monthly Review Press, 1990).

50. Quoted in "State Scan," *Exposure* 36/37 (January/February 1984), 10. Chess made her remarks in consolation of the Texas Environmental Coalition following the defeat of a right-to-know bill in the Texas legislature in 1983.

51. This NIMBY response is understandable in purely economic terms: the redistributive risks and costs of accepting an HWDF often outweigh the benefits to the chosen community.

52. Paul E. Peterson, *City Limits* (Chicago: University of Chicago Press, 1981), 175, generalizes about protest groups: "Analysis of their agitation is instructive in several ways: (1) the divergence of their demands from the traditional patterns of local politics clarifies just how nonredistributive local issues typically are; (2) the processes by which the local system responded to these demands demonstrate the capacities local systems have for handling vigorously stated redistributive demands; and (3) the decline of these groups in local politics shows the fragility of the politics of redistribution in the local arena." These generalizations are apt for protests that call for redistributive policies but not for protests that are resisting redistributive policies, e.g., siting HWDFs. Peterson's frame of reference is generally those protests associated with issues like school financing and housing reform.

53. Ken Geiser, "The Emergence of a National Anti-Toxic Chemical Movement," *Exposure* 27 (February 1983), 3.

54. Ibid.

55. Ibid.

56. Ibid.

57. See especially Adeline G. Levine, *Love Canal: Science, Politics, and People* (Lexington, Mass.: Lexington Books, 1982), and Brown, *Laying Waste*.

58. All of the early newsletters and other publications noted below were written by Lois Gibbs, referring to herself in the third person.

59. CCHW, *Five Years of Progress, 1981–1986* (Arlington, Va.: CCHW, 1986), 6.

60. Ibid., 10–11. The reference to accountability of government and citizen common sense could have come directly from the thought of James Wilson. Cf. Nedelsky, *Private Property*, 96–103.

61. CCHW, *Five Years*, 15–16.

62. Ibid., 19 (emphasis in original).

63. See, for example, Robert Bullard, *Dumping in Dixie: Race, Class and Environmental Quality* (Boulder: Westview Press, 1990).

64. CCHW, *Five Years*, 29.

65. Ibid., 23–24, 42.

66. *Everyone's Backyard* 10 (February 1992), 9.

67. As noted in chapters 5 and 6, HWDF siting is now at a standstill, despite the efforts of federal and state governments to overcome local opposition.

68. See, generally, the "1989 Environmental Justice People's Platform," developed at the

CCHW's Grassroots Convention 1989 and published in the CCHW 1989 *Annual Report* (Arlington, Va.: CCHW, 1989), 10–11.

69. See Bullard, *Dumping in Dixie*, and also GAO, "Siting of Hazardous Waste Landfills and Their Correlation with Racial and Economic Status of Surrounding Communities" (Washington, D.C.: U.S. General Accounting Office, June 1, 1983). Cerrell Associates, under contract with the state of California, produced a study profiling the ideal target community for HWDF siting. The "least resistant" would be "small communities with populations less than 25,000, rural, conservative, low income, above middle age, [whose residents] have high school or less education and nature exploitive [*sic*] occupations such as farming or mining." Penny Newman, "A Mine Is a Terrible Thing for Waste," *Everyone's Backyard* 10 (June 1992), 3. The Cerrell Associates study is referred to frequently in the CCHW's literature as an example of environmental injustice.

70. See, in particular, Gary Cohen and John O'Connor, *Fighting Toxics: A Manual for Protecting Your Family, Community, and Workplace* (Washington, D.C.: Island Press, 1990), and Nicholas Freudenberg, *Not in Our Backyards!: Community Action for Health and the Environment* (New York: Monthly Review Press, 1984). Both of these books present case studies as well as organizing principles and strategies for the grassroots toxics movement. Both offer ample documentation of the connection between the right-to-know movement and the groups supported by the CCHW.

71. On January 16, 1981, less than a week before Carter left office. Eula Bingham, a staunch advocate of the right to know, was soon replaced by Thorne Auchter as head of OSHA.

72. 29 C.F.R. 1910.1200 (1986).

73. These conclusions are drawn not simply from an examination of the standard itself, but also from the remarkably candid admissions of the standard's chief architect, then-Deputy Assistant Secretary of Labor Patrick R. Tyson, who supervised the development of the standard and its enforcement policy. See Patrick R. Tyson, "The Preemptive Effect of the OSHA Hazard Communication Standard on State and Community Right to Know Laws," *Notre Dame Law Review* 62 (1987), 1010.

74. See Brown, "Disputed Knowledge," for

an excellent discussion of hazard communication before passage of the 1983 standard. Hadden, *A Citizen's Right to Know*, offers a thorough discussion of the standard itself.

75. *New Jersey State Chamber of Commerce v. Hughey*, 600 F. Supp. 606 (Dist. N.J., 1985); appealed 774 F.2d 587 (3d Cir. 1985); *United Steelworkers of America v. Auchter*, 763 F.2d 728 (3d Cir. 1985).

76. Hadden, *A Citizen's Right to Know*, 20–24.

77. That accident involved an accidental release of a highly toxic gas used in chemical processing at a Union Carbide plant, a corporation with many plants in the United States. More than two thousand people in the community were killed and one hundred thousand injured.

78. Philip Shabecoff, "Washington Talk: Waste-Plant Inquiry Taps Hot Water," *New York Times*, July 25, 1989, A 20.

79. Jim McNeill, "Protective Instincts at the EPA," *In These Times* 13 (October 11–17, 1989), 8.

80. Associated Press, "E.P.A. Drops Waste Case in North Carolina," *New York Times* (June 3, 1990), A 27.

81. McNeill, "Protective Instincts," 9, and Loughlin, "Waste Triggers Red Flag," *Gainesville Sun*, November 18, 1991, 1A.

82. McNeill, "Protective Instincts," 9, quoting HWTC general counsel David Case. This claim was contradicted by the National Toxics Campaign Fund's report that, in fact, waste reduction among industrial polluters was diminishing the size of the "waste stream."

83. Ibid., 8.

84. Shabecoff, "Washington Talk."

85. See Penelope Canan and George W. Pring, "Studying Strategic Lawsuits against Public Participation," *Law and Society Review* 22 (1988), 385; and for a related legal maneuver, see Carl Tobias, "Environmental Litigation and Rule 11," *William and Mary Law Review* 33 (Winter 1992), 429.

86. Quoted from "They're Here," *Everyone's Backyard* 9 (December 1991), 2.

87. Morone, *The Democratic Wish*, 12–13.

88. Although Title III's passage was hastened by congressional reaction to the Bhopal incident, the structure of the law itself owed much to the right-to-know movement described

in this chapter. The strongest provisions of the bill were drawn from New Jersey's right-to-know law, considered a model for such legislation among leaders of the right-to-know movement.

89. We rely heavily here on Hadden, *A Citizen's Right to Know*, 24–41, a comprehensive, critical account of Title III's meaning for community right to know.

90. In contrast to other portions of Title III, the legislative design of LEPCs was based upon the Chemical Manufacturers' Association Community Awareness and Emergency Response program, a voluntary effort led by industry to establish local emergency planning in the wake of the Bhopal incident.

91. Hadden, *A Citizen's Right to Know*, 16.

8 / Citizenship in the New Regulatory State

1. The terms *centralist* and *decentralist* are from Daniel Press, *Democratic Dilemmas in the Age of Ecology* (Durham: Duke University Press, 1994). His review of the literature convincingly documents the two camps in this debate. That literature is both extensive and subtle, and the conclusion of this book is not the appropriate place for us to engage it in detail.

2. This point is articulated in Robert W. Lake and Lisa Disch, "Structuralist Constraints and Pluralist Contradictions in Hazardous Waste Regulation," *Environment and Planning* 24 (1990), 663–81.

3. Charles Piller reinforces this point explicitly in the conclusion of his study of community reactions to unsafe projects: "Only the adoption of public- and worker-interest standards on health and safety can begin to address the source of Nimbyism. Such standards would dictate caution in projects with unknown effects. In practical terms, greater caution would mean spending the time and money—more than scientists or engineers may consider necessary—to find the best technical approach to limiting environmental hazards or to appraise their health effects. In the long run, however, little would be lost by institutionalizing a go-slow approach designed to build public support." *The Fail-Safe Society: Community Defiance and the End of American Technological*

Optimism (New York: Basic Books, 1991), 196–97.

4. Passage of the North American Free Trade Agreement, with its encouragement of industry flight beyond national borders, complicates this analysis. It is just such a complication, of course, that led to the skepticism with which environmentalists treated the agreement.

5. In fact, the federal courts have consistently been willing to support arguments preferring personal and political freedoms over property rights. See Justice Stone's famous "preferred freedoms" comment in "footnote four" of *U.S. v. Carolene Products Co.* 304 U.S. 144 (1938). It would take only a small jurisdential step for them to sustain a shift in the burden of risk when considering environmental policies.

6. See David M. Trubek, "Environmental Defense, I: Introduction to Interest Group Advocacy," in *Public Interest Law: An Economic and Institutional Analysis*, ed. Burton Weisbrod (Berkeley: University of California Press, 1978), 151–94, for a discussion of the legislative intent of the National Environmental Policy Act. The OSHACT's balance in favor of worker health was upheld in *American Textile Manufacturers Institute v. Donovan*, 425 U.S. 490 (1981).

7. See Stephen Breyer, *Breaking the Vicious Circle: Toward Effective Risk Regulation* (Cambridge: Harvard University Press, 1993), 59.

8. Susan G. Hadden, *A Citizen's Right to Know: Risk Communication and Public Policy* (Boulder: Westview Press, 1989).

9. Benjamin R. Barber, *Strong Democracy* (Berkeley: University of California Press, 1984).

10. For an interesting discussion of similar reforms in Great Britain, see the discussion of experiments in mass political deliberation now taking place in Great Britain, in James S. Fishkin, *Democracy and Deliberation: New Directions for Democratic Reform* (New Haven: Yale University Press, 1991).

11. Karen Witt, "Requiring Students to Volunteer: Learning Experience or Servitude?" *New York Times*, July 29, 1992, A 1.

12. This distinction between public and private is articulated by Hannah Arendt and Jürgen Habermas. For example, drawing upon the work of George Herbert Mead, Habermas also emphasizes the importance of taking the position of the other in personality formation

and the development of communicative competence, *The Theory of Communicative Action*. See also, David Ingram, *Habermas and the Dialectic of Reason* (New Haven: Yale University Press, 1987).

13. In an argument consistent with Barber, Nancy Schwartz calls for a rejuvenation of representative democracy based upon the rejection of "transmission belt" theories of representation. Such theories, which underlie most models of liberal democracy, assume that the goal of democratic institutions is to simply aggregate the unchanged and uninspected preferences of private individuals. She argues that public dialogue is necessary in the choosing of representatives to assure that citizens examine the differences between private preferences and public issues. Otherwise, representatives face an impossible task in representing the interests of the members of their polity: see *The Blue Guitar* (Chicago: University of Chicago Press, 1988).

14. Again, Barber is not sanguine about the uses to which such technologies are likely to be put within existing political/economic structures. Unless institutional reforms are first engineered, information networks and computer technologies place more power in the hands of large corporations and government agencies without significantly increasing the information available to average citizens. For a thoughtful discussion of these issues, one that takes seriously the arguments of communitarian democrats like Barber, see Jeffrey B. Abramson, F. Christopher Arterton, and Gary R. Orren, *The Electronic Commonwealth* (New York: Basic Books, 1988).

15. For an interesting example of how complicated issues can be addressed through repeated and complex public opinion polling that seeks to develop informed opinion, see Peter Neijens, *The Choice Questionnaire* (Amsterdam: Free University Press, 1987).

16. Interestingly, in her reconstruction of representative democracy, Schwartz, *Blue Guitar*, endorses the idea of choosing representatives by lot and examines such practices in early Renaissance Florence.

17. Samuel Bowles and Herbert Gintes, *Schooling in Capitalist America* (New York: Basic Books, 1976), 70, make the same point: "The social relationships of work also affect other areas which, at first glance, would seem to be quite unrelated to work—for instance, the political behavior of individuals. The degree of control over work is a determinant of degree of alienation from, or participation in, the political process. Individuals who have relatively more opportunity to participate in decisions on the job are likely to participate in politics."

18. See, for example, our discussion of the Monsanto case in New Jersey in chapter 6.

19. For details of the Alberta case, see Barry G. Rabe, "When Siting Works, Canada Style," *Journal of Health Politics, Policy, and Law* 17 (1992), 119–42.

20. For our discussion of events in California, we rely on Daniel A. Mazmanian et al., *Breaking Political Gridlock: California's Experiment in Public-Private Cooperation for Hazardous Waste Policy* (Claremont: California Institute of Public Affairs, 1988).

21. Ibid.

22. Rabe, "When Siting Works."

23. Kent E. Portney, "The Dilemma of Democracy in State and Local Environmental Regulation," paper presented at the annual meeting of the Midwest Political Science Association, 1987; and, more recently, see his *Siting Hazardous Waste Treatment Facilities: The* NIMBY *Syndrome* (Westport, Conn.: Auburn House/Greenwood, 1991).

Appendix

1. These data were taken from U.S. Department of Commerce, *Environmental Quality Control*, Bureau of the Census, State and Local Government Special Studies No. 101. (Washington, D.C.: Government Printing Office, 1981). In order to control for skewness and kurtosis, this variable was logged in our regression analysis. We do not control for state size in this variable. We use the actual dollar amount because, regardless of state size, the cost of cleaning up and monitoring hazardous waste sites is constant (e.g., this task costs no less per site in Mississippi than it does in California). In our analysis, however, we do examine the degree to which size of the state budget, rather than the seriousness of the hazardous waste problem, determines state effort in this area. Identifying the size of state budgets as a determinant of regulatory expenditures (rather than

severity of the problem addressed by regulations) is a key component of our theoretical argument; hence, rather than control for one or another of these factors, we make their evaluation the object of our data analysis.

2. We used the sum of capital expenditures and operating expenses for 1980. For Maryland, Nebraska, North Dakota, and Delaware, data for 1980 were unavailable, and 1978 figures were used. For Wyoming, 1977 figures were used. In order to control for skewness and kurtosis, this variable was logged in our regression analysis. The data come from U.S. Department of Commerce, *Pollution Abatement Costs and Expenditures* (Washington, D.C.: Government Printing Office, 1981).

3. We use value-added of these seven industries as the denominator of this variable in order to create a measure of private industry spending that controls for the size of polluting industries in the state. Unlike the case of public spending, there is little theoretical disagreement about the need to control for the size of these industries, when using the figures for private spending. Obviously, the larger the size of polluting industries in a state, the more the private sector spends on pollution controls in that state. That is, private spending on pollution abatement in a state with only one or two small polluting firms will be less than that in a state with several hundred large polluting firms. What we are interested in analyzing here, however, is the interstate variation in spending by private industry once the size of industry has been controlled. We use value-added because it seems to be the most widely used measure of the size of an industry's operation within a particular jurisdiction. These seven industries— Chemical and Allied Products, Primary Metals, Petroleum and Coal Products, Fabricated Metal Products, Electrical and Electrical Equipment, Transportation Equipment, and Paper and Allied Products—were estimated to have produced 92 percent of all the hazardous waste generated in the United States in 1980. See EPA, *Hazardous Waste Generation and Commercial Hazardous Waste Management Capacity: An Assessment* (Washington, D.C.: Government Printing Office, 1980), III-2.

4. U.S. House of Representatives, *Interim Report on Ground Water Contamination:* EPA Oversight, Report from House Committee on Government Operations (Washington, D.C.: Government Printing Office, 1980), 17–18.

5. Council of State Governments, *Book of the States* (Lexington, Ky.: Council of State Governments, 1981), 282–83. In order to control for skewness and kurtosis, this variable was logged in our regression analysis.

6. National Wildlife Foundation, *Toxic Substance Programs in the U.S. States and Territories: How Well Do They Work?*, prepared by Kenneth S. Kamlet (Washington, D.C.: National Wildlife Foundation, 1979).

7. U.S. Department of Commerce, *1978 Annual Survey of Manufacturers* (Bureau of the Census. Washington, D.C.: Government Printing Office, 1981). In order to control for skewness and kurtosis, this variable was logged in our regression analysis.

8. Ibid.

9. Ibid. Ideally, we would use concentration ratios for polluting industries to test these predictions; but such ratios are unavailable for specific industries on a state by state basis.

10. In our early field research, these organizations (and particularly the Sierra Club) emerged as two of the more active environmental groups lobbying for hazardous waste regulation. Membership in these two groups is likely to co-vary with membership in other environmental groups that might be more active in hazardous waste politics in any given state. Our data are compiled from information supplied by the Sierra Club and the Audubon Society. In order to control for skewness and kurtosis, this variable was logged in our regression analysis.

11. John F. Bibby et al., "Parties in State Politics," in *Politics in the American States*, ed. Virginia Gray, Herbert Jacob, and Kenneth Vines, 2d ed. (Boston: Little, Brown, 1983), 66.

12. James P. Lester et al., "A Comparative Perspective on State Hazardous Waste Regulation," in *The Politics of Hazardous Waste Management*, ed. James P. Lester and Ann O'M. Bowman (Durham: Duke University Press, 1983), 216.

13. Both DEMO and COMPET are based upon a weighted average of the following: (1) the average percentage of votes won by each party in gubernatorial elections; (2) the average percentage of seats won by each party in each house of the legislature; (3) the length of time each party

controlled the governorship or the legislature or both; (4) the proportion of time in which control of the governorship and the legislature has been divided between the parties; see ibid., 216.

14. Regression equations were run entering all independent variables simultaneously. Owing to the number of logged variables (rendering actual units of measurement uninterpretable) and to the fact that we are not comparing across populations, we report the standardized beta coefficients. Although we are using a population of the states and not a random sample, we still use significance tests to provide an indication of the magnitude of the relations uncovered in our analysis.

15. Although no single bivariate correlation between independent variables seems high enough to indicate that multicollinearity is a serious problem, the number of relatively high coefficients suggests the possibility of a single variable being highly collinear with several other independent variables. To investigate this possibility, we regressed each independent variable on all other independent variables. See Michael Lewis-Beck, *Applied Regression Analysis* (Beverley Hills: Sage, 1980), 58–62 The R for each of these equations was as follows: NEED = 0.42; EMPLOY = 0.63; ENVGRP = 0.43; ECON = 0.41; STATEXP = 0.50; CONCEN = 0.69; LAWRANK = 0.35. While none of these regressions indicates the presence of a serious multicollinearity problem, the relatively high R for CONCEN and EMPLOY indicates that care should be exercised in the interpretation of their beta coefficients owing to the possibility of unstable estimates.

16. In an additional but unreported analysis, we explored the possibility of interaction effects among LAWRANK, NEED and ENVGRP. We tested for interactions among these three variables, rather than among all the independent variables for two reasons. First, testing for interaction among a large number of independent variables creates a very large number of new variables. Given the limited number of cases, the inclusion of all possible interaction effects would rapidly exhaust degrees of freedom in the analysis. Second, multiplicative interaction terms can create severe problems of multicollinearity, rendering estimates extremely unstable and unreliable. Given the already high intercorrelations among the three indicators of industry strength, we did not feel that we would get reliable estimates from any interaction terms we might create from them. Interaction terms were calculated in the following manner: each variable was converted into z-scores in order to control for their great differences in range; these z-scores were then multiplied to create the interaction terms. In order to control for skewness and kurtosis, these newly created variables were logged. When entered into the regression, none of the interaction terms even approached significance. Further, the beta coefficients of the other independent variables were not appreciably altered by the inclusion of these terms.

17. Further, with the exception of differences in a few variables' significance levels (largely the results of changes in the number of cases), the beta coefficients remain remarkably stable and are consistent with the results presented in table 2. This consistency adds to our confidence that we are tapping true relations among our variables and lessens the likelihood that our findings are the result of idiosyncratic patterns in a small number of outlying states.

Index